Learning Gross

A Book About Your Gross Anatomy Class

by

William C. Forbes III, D.D.S., M.Div.

DORRANCE
PUBLISHING CO
EST. 1920
PITTSBURGH, PENNSYLVANIA 15238

The contents of this work including, but not limited to, the accuracy of events, people, and places depicted; opinions expressed; permission to use previously published materials included; and any advice given or actions advocated are solely the responsibility of the author, who assumes all liability for said work and indemnifies the publisher against any claims stemming from publication of the work.

Dorrance Publishing Co
585 Alpha Drive
Suite 103
Pittsburgh, PA 15238
Visit our website at *www.dorrancebookstore.com*

ISBN: 979-8-88925-393-8
eISBN: 979-8-88925-893-3

To Gerri, author of everything that is good, and wise, and fun in my life.

TABLE OF CONTENTS

PREFACE

For most people, what's inside the body is a great mystery. Haven't you found that to be so? Most people entertain vague ideas about where organs are located, what they do, and how they work. But for those of us who study human gross anatomy, it's different. For those of us who have the splendid opportunity to explore the body for ourselves, to actually see for ourselves the beauty and grace of the human structure, and to learn its secrets in order to better serve our patients, it's a compelling, once-in-a-lifetime revelation. Ours is a uniquely privileged study, and that study would be impossible if it weren't for the unrequitable thoughtfulness of people we've never met, who had the charity to give to us what was their most intimate home for seventy, eighty, ninety years.

We begin that study as strangers – strangers to the human body and strangers to each other – and we invariably finish as good friends to both. I'm happy you haven't missed this rare opportunity. And someday, when you're really old (like me), twenty, thirty, *forty years* after you graduate, when you've achieved your goals as a practicing healthcare professional, you will have occasion to get a whiff of formaldehyde, and you will remember. You'll remember the Anatomy Lab, the names of your lab partners and instructors, and your donor body. And you'll remember what I call the "music" of the Anatomy Laboratory: the sound of learning, the sound of discovery, the sound of students teaching other students, all throughout the room. And you will recall vividly that lovely aroma of formaldehyde.

And here's the thing: when you remember, *you will smile.* I promise you will.

This is a book about your Gross Anatomy class. In formulating it, I've been reminded of a passage in Ecclesiastes 12:12: "But beyond this, my son, be warned: the writing of many books is endless, and excessive devotion to books is wearying to the body."

I agree with King Solomon, the writer of Ecclesiastes. So why am I writing a book? Here's why:

Eva Brann taught in the Great Books curriculum at St. John's College for over half a century. She was the 2005 recipient of the National Humanities Medal and has published various books on the philosophy and poetry of the Greeks. She states, "In learning, as in traveling, and, of course, in lovemaking, all the charm lies in not coming too quickly to the point, but in meandering around awhile." Eva Brann's readings of the great authors come with the easy familiarity of one who has been their constant companion for nearly six decades[1]. Her advice for learning is "meandering around awhile," and so is mine. I've meandered around the writings and drawings of all the great anatomists, explored museums and exhibits, and hung around cadaver laboratories and lecture halls, for more than four decades, probably more years than you've been alive.

Naturally, I'm aware that there is no shortage of excellent Anatomy texts and atlases. We don't really need any more Gross Anatomy textbooks. My book aims at something different from the others. It's about the arduous, challenging, fascinating *process* of gaining understanding of the gross structures of the human body. For most of us who study or teach Anatomy, mastering this process is a lifelong endeavor.

I'm writing what I wish I had known when I was an apprentice to this field. Novices are often advised, "Why don't you write the book that you've always wanted to read, but couldn't find?" As a thoroughgoing anatomist, and an *ongoing perpetual student* of anatomy, this is the book I always wanted to read, but couldn't find. I hope that such a treatise will interest other students and professors of Anatomy. There's no doubt that writing it has interested *me*.

For the student anticipating a cadaver course in Anatomy, there are several books available that were written by people *unfamiliar* with the field, who were privileged to observe an Anatomy course, including the cadaver laboratory component. I have listed four of those books in the "Textbooks and Other Books" section of this book. Those authors wrote eloquently about their experience, with respect, fascination with the subject, and great engagement. However, those books were written by lay people and, in at least one case, a student in the cadaver lab. The book in your hands is different, because it is written by one who has been "professing" Anatomy for many years; it therefore comes with a unique and valuable perspective.

This book is aimed at *learners* of Human Anatomy, whether a student, a teaching assistant, a professor or a curious layman. In my own experience, I've been a perpetual learner of anatomy throughout my long career. Even now, I continue to study the human body robustly. This subject is so *interesting*, so *expansive* that I've become convinced that one can never learn it all. One cannot master everything at once. Consequently, Anatomy professors like myself are constantly committing cruel acts of triage with the information they choose to present to their students.

This book is comprised of notes that I've accumulated over those four decades. I've eavesdropped on other instructors teaching students. I've taken notes during lectures and symposia, collecting epigrams and anecdotes. I've spent endless happy hours in the cadaver lab, with, or without, students. I've spent my leisure time spelunking in anatomical innovations, like plastinations and electronic dissections. Most important of all, I've scratched notes on paper towels after overhearing students talking about what challenged them, and how they overcame those challenges. I have distilled all that anato-mania into this book.

Whether you're beginning to learn Anatomy, or *still learning it*, like me, I think you'll find it interesting. My fondest hope is that you will find it inspiring as well, as you explore this most important subject: *us.*

The "Discourse with the Donor" Sections

Between every two chapters in the book, you'll find an account of an apparition in cadaver form, one of the donor bodies in my cadaver laboratory, who visits me while I'm sleeping and discusses what goes on in the lab. She challenges me, and presents strong opinions about my course. Our conversations shine light on certain components of the laboratory experience, and her pensive perspective gives me a different point of view, much as a doctor presents a second opinion on a patient's health. A fully-functioning anatomy laboratory with 125 students and 30 cadavers has a great number of issues going on; she and I discuss some of them together.

Various Approaches to Our Study

In learning gross human anatomy myself, and in preparing this manuscript, I've explored various approaches to mastering this material. What are the factors that make a student successful? What are some practices that are holding students back? I address these questions throughout the book.

I recognize the tremendous value of the anatomy laboratory, whether it's a lab with plastic models only, or one where students dissect cats and other animals, or best of all, a human cadaver laboratory. The anatomy laboratory's hands-on, 3-dimensional protocols of dissection are the best and most effective learning tools.

I try never to lose sight of the marvelous nature of where I'm privileged to work. The human body is beautiful to see, and astonishing to study. The knowledge we gain applies to each and every one of us, and to those we love and care about. Who wouldn't want to learn it well?

How Anatomy Instructors Think

In the course of their careers, most Anatomy instructors develop various techniques that enable them to learn and retain concepts about the human body. Although they are quick to give the students the information, sometimes they are loath to tell the students their memorizing techniques and mnemonics. That's because we professors like to dazzle the students with our command of the subject and we don't want to give away our secrets. Some instructors don't want to toss around their techniques indiscriminately, presumably because a universal knowledge of those techniques could cause the professors to lose their allure for the student.

This book takes the opposite approach: here are the techniques and learning aids that I've developed, and those I've learned from other professors *and students*, over my many years of teaching and understanding Anatomy. Many of the notes I consider most important are attached to the abbreviation "N.B.:" It means "nota bene," and it means "take special note"[2].

Professors use many kinds of devices to remember the structures and relationships. Be aware that the techniques that your professor embraces go beyond the normal "mnemonic plague" that often afflicts students of Anatomy. When you're struggling to identify human tissue in a cadaver, or on your patient, it's a rare instance in which a mnemonic will help you. The mnemonic will help you know the names of structures, but you need wisdom as well as knowledge, to navigate an actual human body.

Perhaps you've heard the aphorism "*Knowledge is knowing that a tomato is a fruit. Wisdom is not putting one in a fruit salad.*" Now I say to you, "*Knowledge is knowing a mnemonic phrase. Wisdom is remembering how it applies to the human body.*"

So I encourage you to take the approach that your instructor takes: keep your mnemonics and reference plates clearly in your mind as "jumping off places," and then add wisdom to those lists, by pondering each structure: its relationship to surrounding structures, its function, what deficits would occur should it be compromised, those sorts of things.

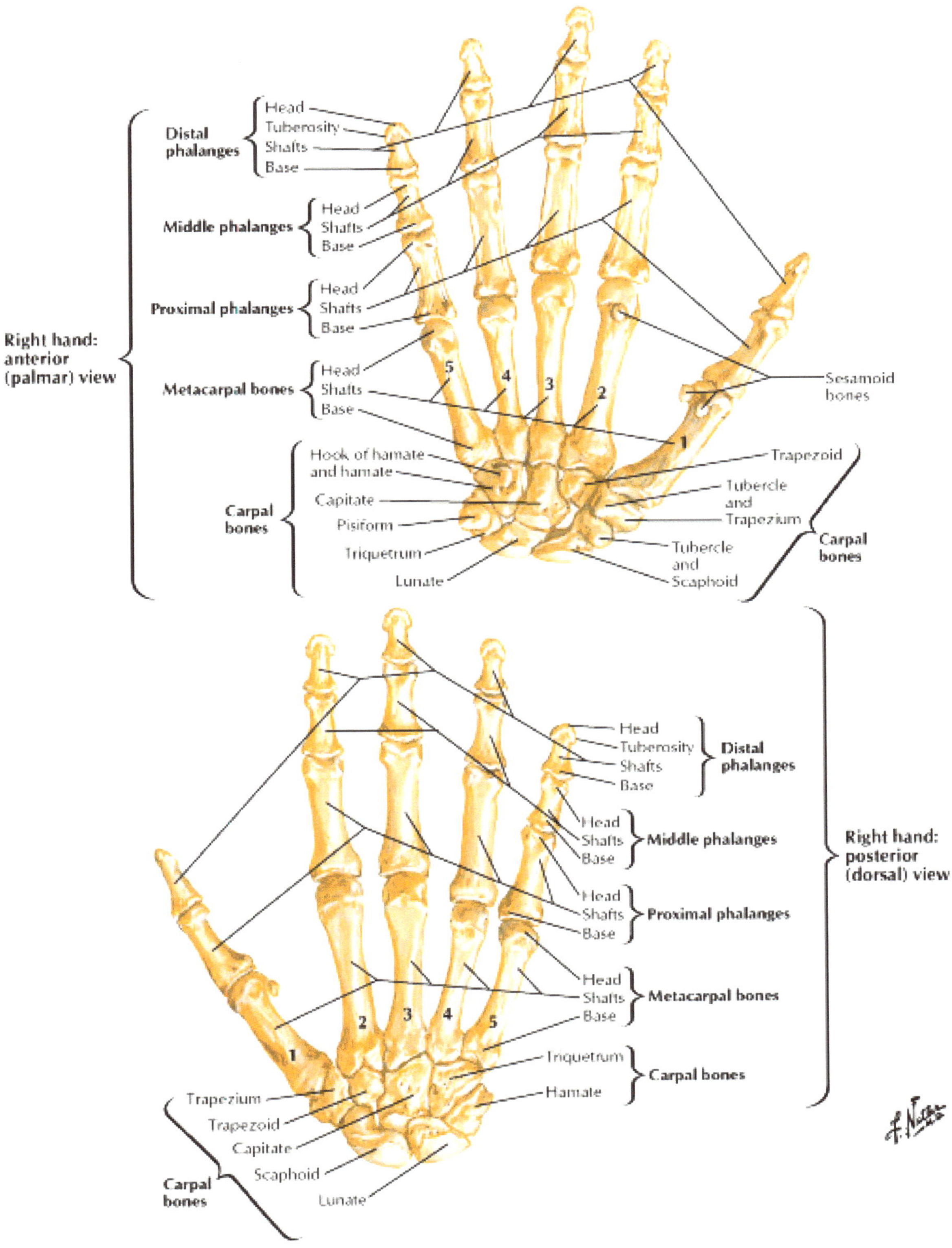

Netter Atlas of Human Anatomy, 5th Edition 444.

How Anatomy Students Think

Ponder this scenario: you've studied hard all weekend, forsaking friend and kin, yet on Monday, when you open the exam, for a fleeting moment you have the vague, uncomfortable feeling that you've come to the wrong room. Your mind is blank. Despite having prepared for this exam with your best effort *all weekend*, you realize that you are going to do poorly.

For example, take a look at the drawing on the previous page, in particular the bones of the wrist, labeled "Carpal bones." You learned the mnemonic that prompts the names of those bones, with all their confusing, nonsensical names. The mnemonic is *Some lovers try positions that they can't handle**. Well done. The problem is this: you neglected learning the *relationships* of those bones, one to another.

*some = scaphoid bone
lovers = lunate bone
try = triquetrum bone
positions = pisiform bone
that = trapezium bone
they = trapezoid bone
can't = capitate bone
handle = hamate bone

Now, when you face the examination where you are required to name a particular bone of a skeleton's wrist, you have a conundrum: using your mnemonic, should you start naming those bones on the medial side of your skeleton's wrist, or the lateral side? Proximal or distal? *Where is* that scaphoid bone that was brought to mind by the first word of the mnemonic? *Where to begin* makes all the difference in the world, and you neglected putting this little detail into your brain.

Consequently, after the exam, you will go to your professor, and you will say, "Dr. Forbes, I'm not a "D" student. I was a merit scholar in high school, I've done well in college, and in preparation for this exam. I gave it my all, last weekend. I'm devoting one hundred and ten percent to this course, and yet I do poorly on your exams. What can I do to improve my score in your class?"

This book is written for you.

How My Non-Anatomy Colleagues Think

My colleagues who teach other biomedical sciences have, on occasion, intimated that my discipline was less difficult than theirs. After all, they tell me, the human anatomy never changes. Therefore, an anatomist doesn't have to continue to read the literature, because nothing new in gross anatomy is being invented or discovered.

This, of course, is codswallop. The approach to *the teaching* of anatomy has changed dramatically since I began many years ago. The *importance* that's placed on various anatomical structures (pancreas comes to mind) has evolved over time. Most importantly, *imaging* has evolved tremendously in recent years, and the person interpreting those images had better have the wisdom of Solomon in Anatomy. Here's an example: my colleague, who is head of Cardiology at a major hospital, told me his residents in Cardiac Surgery don't get nearly the surgical experience they used to get.

"Why is that?" I inquired.

"The advent of imaging in cardiology has virtually eliminated the need for exploratory surgery," he responded. "Nowadays we can get *better* information from our three-dimensional imaging of the living heart than we could get in exploratory surgery. And imaging avoids the trauma of a surgical procedure for the patient."

Of course, this changes everything, including the way we approach teaching Gross Anatomy. Interpreting information derived from exploratory surgery is quite different from gaining that same information on a two-dimensional or three-dimensional image. And imaging continues to improve by leaps and bounds as technology steadily expands.

All that imaging is different from the physical examination, and different, even, from cadaver dissection. But it's one of many advances in medicine which have dramatically increased the need to study the anatomy of a human individual. Anatomy is the floor to many rooms, and the number of those rooms is increasing.

A final point for those who think Anatomy is easier than, say, Physiology.

I think Physiology must be easier than Gross Anatomy. Here's why: in an exam on Physiology, it seems that there are only three possible answers:

1. goes up
2. goes down
3. stays the same.

Dude, given these facts, I could get the first 33% on my Physiology Examination without even studying.

Your Anatomy Resources

In order to take full advantage of your Anatomy course, you should use an excellent Anatomy Atlas. One such atlas is *Netter's Atlas of Human Anatomy*. You can buy a used 5th edition in very good shape on Amazon for twenty bucks. If you use a different edition, use the title of the plate to look up the identical plate in your atlas. *Netter's Atlas of Human Anatomy* is one of the most commonly used atlases in professional schools. I refer to it perpetually, but there are many other excellent atlases out there, as well. In addition, a good medical dictionary is absolutely essential in learning Anatomy. Mine is *Stedman's Medical Dictionary, the 27th edition*, copyright the year 2000 (used, $15.99). Going further, I like to see students use *Gray's Anatomy for Students*. Again, you don't need the latest edition. Got extra money? Buy a set of Acland videos. All of these are described in the chapter entitled *Textbooks and Other Books* (Chapter 13).

Screening What's Important

There are a million facts associated with the discipline called Human Gross Anatomy. You and I cannot learn them all. Therefore, your instructor or professor will triage the particular facts and concepts he or she deems important for you, in order to be successful in the class. You don't get to select which facts and concepts you will learn in their course, and I don't, either. This book cannot, and is not intended to, determine which facts you learn for a particular class. Therefore, you cannot and should not use this book in place of any assignments in your course. Rather, this book is intended to *augment* the resources required by your excellent instructor or professor. It will help you with tips that won't be found in your textbooks, atlases and lab manuals. The purpose of this book is not to *teach you* the anatomy, but rather to enable you to learn anatomy *more easily*, and *better*.

Learn for Your Patients

If you take the attitude that the purpose of your rigorous study is to prepare you to pass your exams, you'll probably get along just fine. However, if you want to make your study immensely more satisfying, try this: *study in order to better serve your patients*. Throughout your career, you will likely be dealing directly with thousands of patients, and you will encounter the need to know aspects of anatomical information that you can't anticipate.

Let's presume that you're a healthcare practitioner, a dentist, say. Do you think the only anatomy you'll be responsible for is *the teeth*? Wrong. Your patients will come for treatment and report that they occasionally suffer from angina pectoris. They're going to assume that you know what that is. Moreover, they will expect you to know how to modify your dental treatment in light of that information. You may even need to answer, for your patient, questions regarding the blood supply to the heart, indeed the *anatomy* of the heart, and when you do, their confidence in your treatment will increase.

No matter what healthcare specialty you pursue, that salutary influence you have on your patients will, in turn, affect many multiples of that number of their family and acquaintances. Your knowledge of Anatomy will have an immense impact on many people. Why not devote yourself to studying with your future patients in mind, and become the very best you can be?

CHAPTER 1:
Introduction to Gross Anatomy

First Day in Anatomy Lab

I arrived at the first session of the Human Anatomy Cadaver Lab with the other instructors, ten minutes before class was scheduled to begin. All the students at my assigned cadaver tables were already present, standing in teams, close to their tables. Their demeanor was serious – even my attempts at lighthearted greetings were met with wan smiles. There was very little conversation, and the conversation that did occur was in hushed tones. I heard one student say that, prior to starting medical school, his mental picture of medical training had been of the Anatomy Lab, and now, "here we are."

Their serious attitudes carried over to the tasks at hand. Before every action, there was discussion among the group: who wants to take the shroud off? Where should we make the first cut? Who wants to be the "reader?" Great deference was shown to the three students (out of 26 in my section) who had experienced a cadaver laboratory before: the members of each group assumed the experienced person should lead.

At the beginning, one or two students at each table timidly stood back, looked over the shoulders of others, and let their teammates proceed with the dissection. After ten minutes or so, they were prompted to join in, by at least two factors: their diminishing discomfort with the situation, and the strong encouragement of their teammates.

The students responded verbally to evidence of the cadaver's humanity: scars, intravenous ports, nail polish, and the like. At this stage, most of them had heard or read discussions about the cadaver being their "first patient," and I saw students solemnly observing the body. When I spoke to them, they indicated that they were reflecting on the fact that this body had been a living person, and what that entailed.

Without prompting, the students themselves instinctively displayed sincere respect for the cadavers. It wasn't necessary to remind them to treat the cadavers with respect; they appeared to bring that concept with them when they came to Lab. I heard no inappropriate speech nor hilarity during this entire session. The first half hour or so, students talked briefly in muted tones, but by the end of the lab session, I began to hear the familiar "music" of anatomy lab: students teaching students, throughout the room.

In closing the shrouds over the cadavers at the end of class, I heard no discussion about who was going to do what, as I had heard at the beginning. The teams had melded, the tasks were not onerous, and the individual students set about cleaning up. They seemed pleased with what had transpired.

The students washing up after class revealed their sense of accomplishment. I heard talk like "How'd it go?" and "Did you guys finish the dissection?" and so forth. In the course of three hours, the demeanor of the individual students seemed to have changed entirely, from "Am I going to be able to go through with it?" to "I can do this," and in some cases, "This is going to be *very interesting*."

Welcome to Gross Anatomy

Since you've picked up this book, I assume you are anticipating taking or teaching a course in Gross Anatomy. In either case, I assume that you've already achieved a great deal in order to get where you are today, and you've earned my respect. I'm delighted that you're interested in this fascinating subject.

Further, I assume you're interested in pursuing a health profession of some sort. I congratulate you on your choice of careers. You join most of civilization in the belief that the best life is the one that is spent in service to others, and that altruism is the loftiest motivation of any health professional.

Let me start by giving you a short overview of a typical course in Anatomy.

The Gross Anatomy Lecture

Most courses in Gross Anatomy involve several essential components: a series of lectures, in most cases followed by lab periods of approximately three hours. The lectures are usually entertaining, and they're seldom jejune. How could they be jejune? *This study, after all, is about us.* All areas of the human experience are influenced by Anatomy, even the most private, personal ones. The lectures in Anatomy will almost always find some application to our lives, whether it's Aunt Matilda's hiatal hernia, or Uncle Ralph's Bell's palsy, or the pain in your belly because it's time for lunch. The study of Gross Anatomy is remarkably applicable to everyday life, and death.

If there is a cadaver experience as part of your Anatomy class, some schools will immediately have their students dissect the area of the body that has just been lectured. Other schools will give the lecture, then give the students a few days to "digest" the material, after which the students dissect the structures discussed in the lecture. The rationale is they can do better dissections if they have heard the lecture *and then studied the material* prior to entering the lab.

Lectures in Gross Anatomy are usually given in one-hour (or fifty minute) sessions. Since Anatomy is so abundantly about relationships between structures, PowerPoint presentations are a common format. Students vary greatly in their approaches to the lectures. Some students boot up the PowerPoint presentation on their laptops and make notes directly on the slide, others take hand-written notes on paper. Some students prefer not to take notes; instead, they focus all their attention on the lecture, planning to make their notes from Panopto or other streaming method later on, at home.

Questions from the audience are encouraged during the lecture. Most lectures include good-natured give and take between professor and students. Some lecturers utilize "clickers," with which they "pop" a quiz question or two in the middle of the lecture, to keep the student's attention on tiptoe. In addition to PowerPoints, the professor may use skeletons, models, Elmo projectors and whiteboard drawings to augment the points he or she is making.

It might surprise you to learn that the other Anatomy professors and staff from the Department routinely attend the Anatomy lectures. We faculty enjoy the review, we "keep the lecturer honest," and invariably we learn something new.

The lecturer does not discuss all aspects of the lecture topic. There is simply not enough time to say all that he or she expects the students to know. Therefore, most lecturers augment their lecture by assigning pages in a textbook, or distributing a notespak, or recommending a video or website.

N.B.: It is my experience that the lecturer will discuss primarily the areas that he or she deems most important, or most obscure, or most difficult. Therefore, while anything assigned is "fair game" for a quiz or test, the student should pay particular attention to the information presented in the lecture.

In reading the professor's notespak, be aware that the abbreviation "N.B.:" comes from the Latin words *nota bene*, and it means *note well.* That abbreviation indicates that what follows is important to that professor. This is an important consideration, and we'll discuss it further in the section entitled, "Know Your Professor," on page 166.

What the Anatomy Laboratory is All About

If you take a course in Gross Anatomy, you'll be anticipating an anatomy laboratory of some sort. I hope you will have had the opportunity to visit the lab in session before you begin your class. Furthermore, I hope your visit comes not at the beginning of the course you're visiting, but in the middle, at a time when the students are accustomed to practicing their art, so you can actually get the Anatomy Laboratory experience.

If you happen to come in the middle of the semester to my former medical school, you will be asked to sign in at the desk. There's a uniformed lady behind the desk who'll ask you to sign your name. Then you'll need to turn to the right, pull on a metal door that's incredibly difficult to open, and pass down the dark and narrow concrete staircase to the basement of the building. (Anatomy Labs are traditionally located in the basements of buildings.) At the bottom of the stairs, you'll pass through another heavy metal door, and you'll see on your left some double doors with the sign stating "No Unauthorized Entry." There, you'll pause for a moment.

It's silent in the hallway. The walls are cluttered with a variety of paintings and pictures of sculptures that might interest someone studying the human body. An occasional student, dressed in the obligatory scrubs, scurries silently down the hall. The sound of doors closing echoes in the empty passageway.

And then you will open the double doors and pass through them into the Anatomy Laboratory. At that moment, a dramatic change occurs. In the hall, there was solemn silence, but as you enter the immense Anatomy Laboratory, you'll be surrounded by a beehive of noisy activity. Students in groups of five or six are gathered around tables, on which there are cadavers in various stages of dissection. Whiteboards line all the walls opposite the cadaver tables, and there are indiscernible scribblings on those whiteboards. Loud conversation pervades the room. Students teaching students. A cacophony of teaching and learning. As a professor for thirty years, when I hear the sound of students teaching other students, I consider it nothing short of symphonic.

Now you should sidle up close to one of the tables, to see and hear what is going on. You'll hear students doing the equivalent of a singer practicing musical scales: moving up and down the structures being studied, calling out the names of muscles, bones, nerves and blood vessels. Repetition has the same value for the anatomy student that it has for the singer: practice makes perfect.

And you will encounter the Anatomy Lab smell, so familiar to anyone who has studied the subject. Years later, these students, when they encounter the smell of formaldehyde, immediately will be taken back in their minds to their first semester of medical school. The same is true for dental students, who are working at the school of dentistry. And the same is true for those studying to be physicians' assistants, nurses and nurse practitioners, physical therapists, occupational therapists, and for students in all the health professions, all over the city, and everywhere.

There is an anatomical reason why the smell of formaldehyde prompts us to reminiscence about anatomy lab: the olfactory cortex is the part of the brain that is associated with our sense of smell. The olfactory cortex is part of an area of the brain called the *rhinencephalon*. The word "rhinencephalon" means "nose brain." The rhinencephalon connects intimately to the area of the brain that controls our emotions, the *limbic system*. Because of the connection of the rhinencephalon to the limbic system, smells often trigger memories and emotions. For students of anatomy, memories of the anatomy laboratory will forever be prompted by the smell of formaldehyde.

The anatomy laboratory is the universal experience of most healthcare professionals. That is for good reason: there is simply no substitute for the touch and feel of the tissue, nor for the three-dimensional exposure to the human body that those students are experiencing. It is the framework on which they will hang the entire study of the structure and function of the human body. Anatomy is the floor of many rooms; those rooms are called pathology, physiology, embryology and many others. *Anatomy omnia vincit.*

There is some variety in this exposure to Anatomy: dental students will spend an entire semester in meticulous dissection of the head and neck. Those students who are planning to be general surgeons will add a more intensive study of the entire body after graduation, during their residencies. Other residencies will offer an intensive, exhaustive dissection in various areas: joints for the orthopedists, skin for the dermatologists, and the female "private parts" for those aspiring to become Ob-Gyn docs. But the initial experience, for all those in most health professions, is to study the entire body: to put scalpel to skin, open it up, and take advantage of a once-in-a-lifetime opportunity to see *what's inside of us*.

If you stand close to one of the groups in the Anatomy Laboratory, you will find that one member of the group has been designated as "reader," a position that rotates through the team every week or so. That person reads aloud from a dissection manual while the rest of the group dissects. Another student may be called "the runner," tasked with learning the prosections (cadavers or parts of cadavers that have been dissected beforehand, usually by faculty or staff). Particular prosections have been set out by the faculty because they are relevant to the area of the body the group is studying. The runner may bring the prosections to the dissection table, or may lead the group to the prosection table to demonstrate what he or she has learned. Other members of the group may take different ad hoc responsibilities: writing and drawing on the whiteboard, or prompting the group with images from the atlas.

Over the course of the semester, as they work together, an easy camaraderie develops among the members of a cadaver table team. In most cases, friendships develop that last throughout school and into their professional careers. One member of my dissection team in professional school went on to academics and became a dean; twenty years after we graduated, he hired me to teach at his school. These friendships are strong, and they are valuable, and they are fortified by the practice of taking a dead person's body apart.

At this particular school, Anatomy dominates the first semester of medical school. Students study Microscopic Anatomy (Histology) in the mornings, and Gross Anatomy in the afternoons. Embryology is sprinkled in. Throughout the autumn, those medical students are completely immersed in Anatomy. When they get out of bed in the morning, they're thinking Anatomy, and when they get back into bed, their thoughts are the same. If they take time out for a movie with their classmates, they'll be talking Anatomy in the theater as they wait for the film to begin. For four months, they swim in Anatomy, and they learn it exceptionally well.

Two students of mine formed the ultimate life-long friendship while in the Anatomy Lab. They fell hopelessly in love with each other in the presence of a dead person, and eventually they got married. Nowadays they practice pediatric dentistry together. Beware: this could happen *to you.*

One more thing about the lab: if you hold reservations about dissecting a human cadaver, remember this: for the timid dissector, dissecting is like getting in a lake in northern Maine: you don't just jump off the dock into the water; it's too shocking. You walk in slowly, and allow your body to accommodate to the temperature of the incredibly cold water. If you enter the water gradually, by the time your body is immersed in the water, you'll find the temperature quite pleasant. "Refreshing" is what we like to say. It's the same in the Anatomy Laboratory.

The Bottom Line

Steve Giegerich entitled his excellent book *Body of Knowledge: One Semester of Gross Anatomy, the Gateway to Becoming A Doctor*[1]. I consider the study of Gross Anatomy to be nothing less than that: the gateway to becoming a health professional.

Please understand this: when you finish the course; when you walk out of the Anatomy Laboratory after your very last session in my Lab, the odds are very good that you will look me in the eye and tell me, "I loved it. I *really loved it and I never thought I would.*"

What's in me's in you. Isn't that amazing? Now let's get busy and find out what that is.

CHAPTER 1A:
Discourse with the Donor

When I started teaching Gross Anatomy, I wondered what a cadaver would think about a cadaver lab if a cadaver could think. Perhaps I spent too much time pondering that fantasy, because nowadays Cadaver Donor Number 2436, who lies on Table 6 in my laboratory, occasionally interrupts my sleep, and, believe me, she is not loath to tell me what she thinks.

"Sorry to wake you," she said, in her hoarse, whispery voice. "I've been thinking about what you said in Lab yesterday, and it makes me irritable, and ugly."

"You aren't ugly. But do you know what *time it is*?"

"You put your hands on my love handles and you said you were going to the gym after class. You were standing at my table, and you made a crack like that. It was offensive. You should have more respect for the people in that room."

I sensed that she was not referring to the living people in that room. "It had nothing to do with you personally. I have the greatest respect for you. Leave me alone."

"I may lack a functioning limbic system, but I've got feelings," she complained. "You wouldn't care about this, but I was very attractive when I was alive. When I had hair. Before that jerk shaved me like a billiard ball." She gently ran her fingers over her cranium and walked around the bed. She pointed her finger at me, the skin of her palm flayed and flopping. "Do you know who I was?"

"I don't know much about you, actually. I know you have some very expensive dental work, and your fingernails have pink polish. I know you've had bypass surgery but not a hysterectomy. I know you still have your wedding ring, apparently because the coroner couldn't remove it over the knuckle. So I know you were married. I do know a little about you."

"Oohhh, I was married, all right." The left side of her face, the side that was not yet dissected, smiled demurely. "He was so devoted to me that all my friends were jealous. I had this special control over him, that nobody understood, not even him."

"I don't want to talk about it," I said. "Your sex life is no concern of mine."

She jumped immediately on that. "Oh is that so? I beg to differ. You've shown my uterus to dozens of people. You've described the fertilization process and demonstrated my Fallopian tubes to the world! I'm so tired of that. It's *so* tedious." She turned and walked toward the door. I hoped she would leave, because I was becoming uncomfortable with the way the conversation was going.

She came back. "And another thing," she said sternly, her lips pursed like a schoolmarm, "I don't like my private parts laid open like a library dictionary. I don't particularly like my *sitz-fleisch* exposed in public. It's the same with the male bodies: their infernal priapism is discomfiting. Just throw a towel over those places when you're not dealing with them."

"I'm sorry. But what did you expect? We need to study the entire body, and that's part of it." I tried a compliment. "Most of the female cadavers don't have their uterus nor their ovaries, so you're special in that way." It was a weak, obvious attempt to appease. I held back what I was going to say: that we planned to sagittally section her entire pelvis soon.

"I know about the sectioning." She had read my mind. Or maybe she had read my face. "I'm O.K. with it, I suppose." She looked pensive. "After all, I'm not exactly a Netter's Atlas model, am I?"

"Grant's Atlas models are better, anyway. They wear a serene, peaceful look. That's what I want for you," I said.

She seemed to appreciate that. "I worked hard, when I was living, at keeping this body healthy. I was a runner, until my knees played out. They went varus. But I suppose being varus helped when I had my kids."

She began to pace, and her one and only breast swung side to side as she did so. Her voice was soft. "I miss those days," she said.

I changed the subject. "Are you afraid of cremation?" My reactions to her were indescribable. I detected in her eyes a mysterious world which is not open to me.

"Everybody's got to go sometime," she sighed. "I knew about the cremation when I signed up. Just be sure you don't waste this," she said, peering down over her open abdomen. "Just be sure you do a good job dissecting this. This body did everything I asked of it for sixty-two years, and I expect it to be treated with respect." She turned to leave.

"I'll see to it," I said.

CHAPTER 2: Relationships with Professors

Your Professors Want to Like You

I can't stress this enough: *get to know your professors.* When you see them in the hall, look them in the eye, and greet them by name. When they give a good lecture, compliment them for a job well done.

I don't mean schmoozing. I mean be friendly, cognizant of the fact that your career can be enhanced by a good, collegial relationship with them. Don't kiss up. Just be nice. You might be pleasantly surprised at the professor's response. Most professors want to like their students. However, they get annoyed when their only personal interaction with a student is when that student shows up in their office with some jeremiad about how hard they're studying. Don't be that person. It could affect your grade adversely, and it's a lousy way to be. If you have a genuine problem, or a question about the material being presented, go see the professor in his or her office, during office hours. Get your questions answered; be respectful of their time; don't stay too long.

If you're in a class of 300 students, it may be unrealistic to get to know the professor. In that instance, probably the class is being taught by a TA anyway. But in professional school, or graduate school, the student/professor ratio is more salutary. There, your professors are the engine that drives your academic success. Have an adult conversation with your professor, discuss the topic that's giving you trouble, and then be sure to respond appropriately to non-verbal cues that he or she needs to get back to work.

If possible, become acquainted with one or two professors in particular. These should be professors who are interested in what interests you. I teach Anatomy, and students whom I've gotten to know over the years are often those who plan careers in surgery, obstetrics, sports medicine, dentistry, physical medicine, and the like. We have a great deal to talk about.

A professor who's acquainted with you through your student years is in a position to introduce you strategically to his or her colleagues and friends. Your professor friend will also write letters of recommendation for you, give you advice, and help you take advantage of the resources that are available to you.

Succeeding in Your Relationship

My son Bill achieved a fancy GPA at college, and I asked him how he did it. "I psyched out my professors," he said, simply. When I pressed him to be more specific, he said, "I tried to figure out what components of the course were most important to them, and I concentrated my study on those areas more than the others." I think this is an excellent approach. One way to figure out which components of the course are important to the professor is to attend the lectures *in person.*

Another way to determine which components of the course are important to the professor is to visit the professor in his or her office. Most professors post office hours, where they meet with students on a regular basis. That's the best time to stop by the professor's office. Other times can be arranged by phone, email or text. By visiting his or her study you'll begin to develop rapport, and your friendship, like any friendship, will grow. He or she may become your mentor, who deals not only with your concerns and questions about the course, but also gets to know who you are, what you're about, and your career plans. Your friendship will motivate you to achieve, and not only for yourself. When you ace that exam, one of the first people you'll share the achievement with is that professor.

Dude: if you visit your professor in his/her office to clarify a point, *be prepared* by studying the material before you go. Go over the lecture or lab that you've issues about, organize your thoughts before you go, avoid maundering, and get straight to the point. If you don't do that, and if you show up unprepared, expecting the professor to spend his valuable time *re-teaching you* the material that he or she has already lectured about, be sure to take doughnuts. Your professor's irritation is not pedagogical petulance; his or her disappointment results from the fact that we professors are perpetually short on time, and we abominate wasting it. When you've shown up in his or her office unprepared, you've made an adverse impression on that professor, and now that toothpaste is out of the tube.

When you visit your professor, don't be chummy. Some students, educated in an era when students address their high school teachers by their first name, address me similarly, in an odious attempt to efface a chronological interval in our ages of nearly half a century. That interval is deserving of respect, and it is a clear mistake to neglect opportunities to show deference to your professors.

Your respect may be unrequited, and may even be met with condescension. It might help you to know that some professors practice the teachers' maxim "Don't smile until October," intending to start out being stern, and plan to soften up later on. Whatever their demeanor, you must always address your professors as "Doctor So-and-so," or "Mister (or Missus) So-and-so." (Be sure you know which title is apropos.) Address your professor in these terms, but do so in a manner not emphasizing his or her relative senescence. Some of us professors are quite superannuated, a situation that is, anatomically speaking, evident in the gradual diminution of our cervicomental angle (due to the southward drift of our hyoid bones). You should find this maturity worthy of respect.

When you go to see your professor, prepare, with a specific question or questions to ask. You'll find that when you ask a question, you carve a place in your brain for the answer, and you'll remember that information longer than if you simply encounter it while reading a book. The fact that you have a query doesn't make you a weak student; in most cases the opposite is true. You *must* ask the question so that you can learn the answer. Remember: you may actually *see* that question you didn't ask, on your next examination, so be sure you know the answer. Just *do it:* ask the question.

Furthermore, in your conversations with your professor, if you're not completely clear, try this: play back what you think you heard from him or her. Try to teach it to the professor. He or she will immediately identify flaws in your thinking or errors in your information. Don't be afraid to echo what you heard and then to be corrected.

Even the vaunted Scientific Method begins with a well-thought-out question. It starts with a hypothesis. It follows up with systematic observation, measurement, and experiment, followed by the formulation, testing and modification of the original hypothesis.[1] The beginning of learning is the question.

N.B.: When you come to my study, please don't waste my time asking questions to which you already know the answer. Believe me: we can tell. Here's the point: *we professors expect from you the same things that you expect from us*: be prepared in advance, pay attention to what is going on, give it 100% of your effort, and bring a professional attitude to our meeting.

If you need to see your professor, don't wait until the last minute. See him or her before the stuff begins to avalanche. Just like an unresolved issue in your marriage, this problem will only compound over time. It's strange, but sometimes the students' procrastination in coming to see me reminds me of the patient who showed up at his doctor's office:

"Where have you been?" the doctor asked. "It's been many months since you've come to see me."

"Oh, Doctor, I couldn't come because I've been sick," the patient replied.

It's possible to wait too long, on this. I can't count the number of times a student waits until he or she is in a position like a person tipped 'way back in a chair. The chair is going over backwards any second now. Not until that point does this anxiety-laden suffering student decide to come and talk to me. When they do come, we talk about the problem. Here's something I've noticed: in explaining their poor performance, the student often presents me with drawn-out circumlocutory answers that beat all around the issue, instead of getting straight to the point.

You and your professor will discuss your study habits, and he or she will make recommendations. Remember: addressing learning problems, just like addressing medical problems, requires a diagnosis before you can develop an effective treatment plan. So be candid with your professor regarding your study habits, and remember that the professor is in your corner. The professor wants you to succeed. Give him or her the truth about your situation, and consider the advice carefully.

Another tip: don't ask to visit the professor after a three-hour lab session. Many of us are considerably into our dotage, and three-hour laboratory sessions can be fatiguing for us. Most of us just slip into maintenance mode after a long lab. After talking non-stop for three hours, we need time alone, to recuperate. Unless the professor schedules office hours after labs, it's a good idea to make yourself scarce. Check out the office hours, posted on his or her office door.

A related topic: most professors have times when they need to bear down on some task to which they're obligated: an upcoming lecture, a review of structures in preparation for a lab, or a letter they're writing professionally. Professors will often close their door in those instances. You may "stop by" their office, and see that the light is on inside, and in my case, there's a little window down low near the floor where you can actually see into the professor's study. It's courteous, and professional protocol, to leave those closed-door professors uninterrupted. Don't knock. During times when the professor welcomes visitors, he or she will leave the door open. And when it's open just a crack, the invitation is somewhere in between. You will see faculty honoring these protocols with each other, and you should honor it too.

Your Feedback

Those visits with professors should be mutually beneficial; they may ask for an opinion from *you*, as well. If I were the student, I wouldn't give negative feedback, nor would I cavil about the wording of test questions, nor would I propose changes to improve the course, unless it's asked for, *and even then, I'd utilize my greatest diplomacy in giving it.* Remember: professors are defensive about their course, and can be tetchy about receiving criticism, even if it's meant to be constructive. I often say that a professor's course is like one of his or her children: criticism, even the constructive kind, is usually not welcome. Instead, focus on positive aspects of the course.

Even if you had a *really good* Anatomy course previously, save all the helpful tips about how you did it before. If you must make suggestions, save them for your Course Evaluations at the end of the course. You must stifle yourself like a compulsive jokester at a funeral. It reminds me of a common bumper sticker, seen 'way up in rural, woodsy, Northern Maine. It's aimed toward people "from away," who move to Maine for the attractive rural lifestyle, and immediately want to install more traffic lights and a symphony orchestra. The bumper sticker states, "We don't care how you did it back in Boston." Your professors, like the locals in Maine, don't always appreciate your suggestions and critical analysis.

Of course, if he or she asks you a specific question about the course, summon your best diplomacy, and give a frank answer. If your professor's response is abrupt, don't press the issue. Considering their "power-over" position, professors may not be entirely candid with you about their response to your suggestions. They realize that if the conversation turned eristic, you can't fire back. Therefore, professors rarely say anything that might be characterized as a hit-and-run.

Don't be a schmoozer. Be a person of manners, but don't be mannered. It's natural and legitimate to befriend someone who's been through what you're going through, and whose job it is to help you get through it yourself. So go ahead and attend his or her office hours, and when you go, be prepared. Be a friend, and a potential colleague. A mentee. You'll be glad you went. And so will they.

Anatomy Lab Instructors

Notice how your instructors love the Anatomy Lab. We actually *compete* with each other, in some instances, to be allowed to teach in the lab. It's that way with me: I like to lecture, I like meeting with my students, but I *love* the anatomy lab. There's simply no place I'd rather be. Bill Hayes, in his excellent book *The Anatomist*, states that H.V. Carter and Henry Gray, who together wrote the original *Gray's Anatomy*, shared a passion "verging on the macabre," for dissecting[2]. I've known people like that.

I've taught Gross Anatomy in a variety of health professions schools. In my long career, I've taught at two medical schools, two dental schools, a pharmacy school, a nursing program and a nurse practitioner program, a physician's assistant school, and a physical therapy school, a mortuary school, as well as undergraduate courses and several residency programs. Students at these levels are self-motivated, smart, and energetic. You don't get admitted into professional school by being indolent, so teaching professional students is a great gig, and we professors love what we do. Students are *so* motivated in professional schools that my personal philosophy of teaching in a professional school is this: stay three pages ahead of the students in the textbook, and *keep out of their way*. You may be surprised, yourself, at your strong desire to know this stuff.

Here's something that may surprise you: in many professional schools, faculty are discouraged from meeting with small groups of students apart from the regularly scheduled laboratory sessions. Students are offended by this. What is the rationale for this school policy? It's this: when faculty members make a practice of meeting after-hours with students, some of those students, (not you, of course) will "blow off" the regularly scheduled lab sessions, because in those sessions they have to share a professor with several other dissection tables. After disregarding the regular lab period, those students arrange to meet a professor in the lab "to help me see a few structures that I'm struggling with." There, they have the professor all to themselves. In most institutions, this is considered unfair to the students who come to regularly scheduled lab sessions, work on their own, and share the instructor.

The result is that, at my school at least, faculty typically don't meet a student or a small group of students in the lab after hours unless that student has missed lab in an excused absence. I'm the first to recognize that there are two sides to this decision. On the one hand, some say, "We intend to do everything we can to aid the students' learning, so why not meet separately with them in the lab?"

On the other hand, some say, "If we do that, the professors will be spending all their time in the lab, and in the regular sessions we'll see very few students." Resolution of this dilemma is peculiar to your specific institution, and the discussion is ongoing.

Professors and the Struggling Student

Over the course of my career, I've participated in the hiring of a few Anatomy professors. During that process, we Search Committee members have encountered a wide variety of professorial types, whose characters range from avuncular to aloof, and we've been keen to distinguish one's pedagogical philosophy from another's. Ergo, during the interview, we invariably ask: "What do you do with a student who is struggling academically?" In my estimation, there is an incorrect answer, and a correct answer, to this question.

Incorrect answer: if they're struggling, the student is probably in the wrong profession. Not all students are made to be health professionals. If certain students can't cut it academically, we don't do them any favors by allowing them to advance. If you can't stand the heat, get out of the kitchen. Better to nip this problem in the bud, and dismiss them, and the earlier the better.

Correct answer: every health professional knows that you must have a diagnosis before you have a treatment plan. What is the problem that the student is having? Trouble at home? An unsympathetic spouse? Money problems? A health problem? Distraction by a love interest? Let's have a frank conversation with the student, address what's inhibiting his or her success, and see how, together, we can ameliorate the student's situation.

Here's one example: for the academy in the new millennium, there is no greater virtue than diversity. The result is a wonderful rainbow of cultures in most institutions of higher learning. There are some cultures in which men and women, even men and women in rigorous academic programs, are expected nonetheless to continue to fulfill *all of* their domestic duties completely. It's all too common in professional schools today: the student's family doesn't understand or appreciate the rigors of a professional school. The student is handicapped in this instance, and if the faculty and administration are unaware, they may be unsympathetic, and unhelpful to the student. However, if we are acknowledged of the problem, we'll be more than willing to help that student.

Do let your advisor know if there are things going on that affect your study adversely. Then take the advisors advice seriously. *We want all of you to succeed*

Some Lectures Would Put the Fire Out

If I were you, I would always attend Anatomy lectures. I am fully aware that for some of us professors, our fireside chat would put the fire out. You know who I'm talking about: at the end of their lecture you think, "I'm dumber now." I've seen students watching a lecture with all the interest of cows in the field watching a car go by. They could have learned more by staying home and reading the textbook. *Or not.*

I understand that you can watch the lecture on Panopto, later, at double speed in order to save time. Dude, listen: there's still no substitute for being present in the class. Even if it's uncomfortable for you to sit for long periods of time, find ways to deal with it; if you find that you're losing circulation in your gluteus maximus, stand up in the back of the room.

Some schools require students to "swipe in," and the school limits the number of absences. In many professional schools, guest lecturers present periodically, sometimes in a "lunch and learn" format. Maybe they're clinicians in the hospitals, maybe they're hospital administrators, maybe they're politicians looking for your vote. If that lecturer shows up to a class of 100 students, and there are 25 students present, that lecturer is not going to return, and the Dean of your school is not going to appreciate the resulting egg on his or her face. Be respectful. Put yourself ahead in your career 30 years, and consider the lectures you'll be presenting, when the situation will be reversed. In most cases, it's common courtesy to show up.

Jennifer L. Taitz wrote an article in the Wall Street Journal named *The Case Against Cutting Class*[3]. In that article, Dr. Taitz, who is a clinical psychologist, wrote :

> Getting up for lectures each day will help you cultivate good sleep habits, since one way to treat or prevent insomnia is to maintain a set wake time. If you stow your smartphone and mindfully participate, not only will you actually learn, but you may find yourself less stressed than when passively scrolling through social media or frenetically texting. Your British Literature discussion may prove a nice distraction from ruminating about your relationships. And sitting in a room full of people you have something in common with is an opportunity to create meaningful connections and feel less alone.
>
> If you assume sticking to your schedule feels forced – especially if you're tired, hung over or behind on assignments – go anyway! Behavioral activation, or sticking to a meaningful plan independent of your mood, is as effective in treating depression as medication. Even when you don't feel captivated by your required courses, arriving with your eyes up will help you live better and

> prevent the panic and sadness that are bound to torment you if you fall behind or isolate. After you commit to going to class, if your mind comes up with an excuse not to go, see that as mental spam, not a sensible plan."

It's worth considering that if you're bored in the lecture, perhaps the problem is with you, and not with the lecturer. Participate in class. Monitor yourself: am I engaged? Am I paying attention? If you're asking yourself, "Should I ask that question?" you should; some of your classmates are having the same difficulty. Come to class prepared with a creative question to which you genuinely don't know the answer. Be congenial and collegial with your instructor. *They want to like you.* Sit in the front row. If sitting in front is a problem because the instructor spits when he talks, bring an umbrella to class. That's what my students do.

Your Professor Marvels With You

Be sure to give your instructor some form of attestation that you marvel at what we get to study and learn in Anatomy. We professors certainly marvel at it. I'm baffled by students who don't. I have students who can bore me for long periods of time about the wonders of their iPhone, or some new social medium that they've enrolled in, and yet those devices are child's play compared with the body they're living in.

St. Augustine wrote, in his *Confessions*, "Men go abroad to wonder at the heights of mountains, at the huge waves of the sea, at the long courses of the rivers, at the vast compass of the ocean, and they pass by themselves without wondering."[4]

Think about this: if one were to stretch the DNA molecule out, it would be 6 feet long, yet every one of the 100 trillion cells in our body has it. There are three *billion* letters in the genetic code. All three billion are faithfully copied in every division of virtually every cell, in exactly the right sequence. This miracle is the essence of the astonishing body that we get to study. We Anatomy professors have spent our lives learning the human body, and we still find ourselves perpetually amazed. Don't be a sufferhead. Go ahead: allow yourself to be amazed with us.

CHAPTER 2A:
Discourse with the Donor

Somehow, through those mysterious and inscrutable corridors and passageways of cadaver existence, she arrived at my bedside carrying a small paper bag, obviously weighed down with heavy contents. Gradually waking from a deep sleep, I was *non compos mentis*, but when I saw the bag, I immediately became alert, and curious.

"What's in the sack?"

"Dirty dissection instruments. Body parts."

"Hmmm… nice. My students bring me donuts and coffee."

"This is not funny. This is the stuff that students leave strewn all over the place when they leave the lab. Some of these were even left inside the shroud. There are scalpels, and scissors and sharp probes, here." She sifted through the articles with the naked bones of her fingers. "They leave the lab looking like a trailer park after a tornado. If this continues, Habibi, somebody is going to get hurt."

"You brought me dirty instruments at…" I glanced at the clock. "…2:31 AM? O.K., I'll try to do better. Now leave me alone, and let me get back to sleep."

She ignored that request. "It's a disgrace, the mess they leave. Their instruments lying around everywhere. And then there's *this. Look at this!*" She held out a nasal septum, dangling it from two of her finger-bones away from her body, the way you'd hold a snapping turtle. "They'll be looking for these when they prepare for the midterm, professor." I had no response; I was in complete agreement.

I said, "I think sometimes the students don't know where they can put those small body parts when they leave the lab."

"Too bad they can't ask *me* where they can shove those items," she said, with a conspiratorial smile. "I'd be happy to suggest a suitable repository for them." She paused for a moment. "Aren't they supposed to wash the instruments at the end of a lab session?"

"Actually, no. They can put them away dirty, if they want to. We don't have a policy about that. They just need to put them away. Sometimes, when they spend their lunch break in the lab, they don't clean up properly when they leave."

"Don't the dirty instruments make the Lab smell worse than it already does?"

I've always treated the smell of an anatomy lab with insouciance. I said, "I don't think they contribute to a bad smell in the lab. Besides, I don't mind the smell. In fact, when my wife and I go on a date, I have her splash a little formaldehyde behind her ears. It's a real 'turn-on' for me."

"I suppose that's an acquired taste," she said with a smirk.

I went on, "I've written specific instructions about care of instruments and body parts, on the whiteboard at the front of the room. I think that's enough."

"Yes, and your handwriting looks like that of a four-year-old boy, raised by wolves, who just learned how to hold a crayon. Doctor, you've got to make it clear to them: these instructions are not *suggestions*, they're compulsory."

"I suppose you're right," I said. "But I hate getting on their case. I want to be more welcoming to the students. To tell the truth, I like the idea that the anatomy lab is a 'clubhouse' for those students who are interested in Anatomy. I like it when groups of kids hang out there between classes."

"Hmmm…" she said. "And in the process, they just might learn a little anatomy. I agree with you." She reflected a moment, and then suddenly she said, "Except when they play music on their phones. I get so *sick* of hearing… *Tenacious Fungus Band*, every noontime. It's musical ipecac!"

"You should relax about that. The Apostle Paul said, 'For ye suffer fools gladly, seeing you yourselves are wise.' Perhaps you should aspire to be more like Paul." I was proud to show off my knowledge of the Bible, knowing she was inclined that way.

"I don't suffer fools, gladly or otherwise. Never have. The music the kids play in the lab is insufferable."

We were silently in agreement, so I asked the question I'd been pondering all evening. "What was the crisis at Table 5 today?"

"One student accused another of using hurtful phraseology. The offending student had observed, with a twinge of envy, that his donor body had accoutrements indicating a successful life: fine dental restorations, a wedding ring line, breast implants and skin that belied her age, which was 92."

She stood there, Specimen 2436, Table 6, inventorying her own crowns and bridges with the remaining half of her tongue while I asked, "Why were they arguing about that?"

"Marie is just such a scold. Rex offended her because of what he said. The gravamen was that his remarks indicated an admiration for evidence of privilege."

"*Privilege!*" I felt my eyebrows rise almost to my hairline. "Seriously? A cadaver? She got the vapors over *a dead body's privilege*?"

The wise cadaver considered. "I rather think she has a point. I don't walk in her shoes. At the same time, it makes me weary when these busybodies constantly have their antennae tuned towards micro-aggressions and pseudo-offenses. All this kvetching is childish, and it's unprofessional."

I felt it necessary to change the subject. I looked at her sinewy body and said, "You look like you were quite fit when you were alive."

She wore an expression one would wear when shedding a tear, but in the absence of lacrimal glands, it was simply a facial expression. Her eyes were black, like the night sky when there is no moon. "I was. I was very skinny in life. My husband joked that if I hadn't had an Adam's apple, I wouldn't have had any figure at all. I almost always wore a sweater. That was in another lifetime. Then I got *old*, and I experienced this inevitable decrepitude that comes with ageing." It was quiet in my bedroom for a while. "Now," she said as she looked down at her rectus abdominis muscle, dissected and reflected to expose beautifully the superior and inferior epigastric vessels "…I hate all this… *necrosis*."

"Well, I think you're beautiful," I said.

"Why, thank you. I'm certainly more attractive than that fellow on table seven," she said. "He's a big guy. And he has a splendid panniculus adiposus." She chuckled, as she recalled his arrival to the lab. "The shroud was wrapped so closely around his body that he looked like he was wrapped in butcher paper."

"But sometimes those bigger bodies are better for dissection," I said. "If you're willing to put the time in, they have big muscles and bones, and their structures are easier to locate."

We pondered for a moment the arduous process of removing thick adipose tissue, and then the happy discovery of strong, easily identifiable muscles underneath. "More work, less drama," I said. "We need to remove his pacemaker before the course is over," I said, reminding myself. "Pacemakers can explode during cremation, and we don't want that."

"Well, have a wonderful time with those big muscles, *Doctor.*" She had emphasized the last two syllables, like a nurse to a senile patient. "If you're looking for perfection, I'll be on Table 6. And," she whispered conspiratorially, "I promise I won't blow up the morgue." She wore the look of contentment and pleasure that cats have when they are purring. She squeezed my shoulder with surprising strength, considering the students had removed the flexor digitorum muscles from her forearm. With that, my friend was gone.

CHAPTER 3:
Anatomical Terminology, Etymology and Abbreviations

The Unfamiliar Word
"Would you think me absurd
If I used an unfamiliar word?
Would you think me a show-off;
Someone you would blow off?
Even if a word is abstruse,
Don't be timid about its use:
Correct manners are good
Being "mannered" is acting not the way you should.
What if I were to deduct
All use of the word "usufruct?"
Unfamiliar words are eschewed
And using them's considered rude
But sometimes the word is a gem
It's just unfamiliar *to them*."

WCF

Your Parents Were Right
Remember when you were in high school and your parents attempted to persuade you to take Latin? *Why would I want to take Latin? Everyone knows, it's a dead language. Nobody even speaks it anymore!*

Sit down and take a deep breath. Your parents were right. Knowledge of Latin can help you learn almost anything, and nowhere will it be more valuable than in your study of Anatomy.

Gross Anatomy has the reputation for being tough, and one of the reasons that it's tough is because it involves learning names and descriptions of a zillion structures. If you learn those names and descriptions in the same manner you would memorize nonsense syllables, you will struggle in Anatomy. On the other hand, if you break the words down, and if you consider carefully the Latin or Greek derivations of each part, you'll learn them more easily. And you'll recall them more easily, later on. If you know what the parts mean, you'll find that *res ipsa loquitor*. (Latin: "The thing speaks for itself.")

For example, here's a word with which you may find Latin analysis helpful:

Callipygian: having well shaped buttocks. From the Greek word *kallos = beauty*, and *puge = buttocks*.[1] If you happen to be callipygian yourself, I suppose that's a good thing. If not, you can work on that. Gluteus major is strengthened by running, but not so much by standing and walking. Gluteus medius and gluteus minimus are more involved with walking. And this part is counterintuitive: the importance of those latter muscles in walking is not in simply moving alternating lower extremities posteriorly; gluteus medius, for example, keeps the pelvis on one side from sagging, when you lift off the ground the foot that's been supporting it.

Here's a cute etymological example: the pudendal nerve and the internal pudendal artery serve the external genitalia and the structures located where your lower extremities come together. We Presbyterians refer to these structures as "the private parts." Evidently early anatomists agreed: the word "pudendal" comes from the word "pudeo," meaning "to feel ashamed."[2] Knowing these quaint facts helps me to recall this word, in all of its various forms.

Every language has a grammar*: a set of rules that govern usage and meaning. The language of Gross Anatomy is no different. A huge part of mastering any kind of Anatomy is learning the language.

Brace yourself: here's another area where a parent (in my case, my own mother) was correct: if you use a word properly three times, it's yours for life. Absolutely correct.

* Grammar: the whole system and structure of a language or of languages in general, usually taken as consisting of syntax and morphology (including inflections) and sometimes also phonology and semantics.[3]

Anatomical Terminology

One of the first things you'll learn in your Anatomy course is the anatomical terminology. Of course, you'll want to have the terms firmly in your mind. Throughout the course, certain words will be used quickly and casually by your professors and classmates; you'd better know the difference between a sagittal plane and a midsagittal plane without too much contemplation. (Midsagittal plane divides the body into right and left halves, and is *right down the middle* of the body, while an unlimited number of sagittal planes are parallel to it.)

What is the provenance of this terminology? The names of anatomical structures are almost always derived from Latin or Greek. You really don't have to know whether the word was derived from Latin or from Greek, in order to benefit from the practice of learning the derivation of the word. In the dictionary:

(L) means Latin

(G) means Greek.

Who decides that the vein in the shoulder should be named "cephalic vein?" And why are the terms anatomists use constantly changing? If you employ an excellent medical dictionary in the course of your studies, you will encounter, associated with certain words, the abbreviation "[TA]." It stands for *Terminologia Anatomica*, which is defined by my medical dictionary as "a system of anatomical nomenclature, consisting of about 7500 terms, devised and approved by the International Federation of Associations of Anatomists (IFAA) and promulgated in August 1997, at Sao Paulo, Brazil."

The definition goes on to say, "Since its foundation in 1903 the IFAA has held periodic conventions for the standardization of anatomical concepts and terminology. In 1989, the federation elected a 12-member Federative Committee on Anatomical Terminology (FCAT), consisting of experts from 11 countries… With the election of additional members in 1994, the FCAT had representatives from 16 countries and 5 continents. The committee solicited suggestions from anatomists and others from around the world, and from more than 10,000 terms… they formulated a list of those deemed worthy of consideration. During eight years of deliberations they chose the simplest and most exact terms, preferring those that are descriptive of form or function over semantically opaque ones. Some 10% of formerly accepted terms were rejected or altered because they were considered inaccurate, ambiguous or otherwise unacceptable… Adoption of the new terminology is expected to be widespread. Because English is spoken in many countries and serves as a common language for scientific and medical communication, English equivalents of Latin terms are given in the published version of TA. However, only Latin terms have official status…"[4]

The student of Anatomy soon learns to be precise in his or her terminology. Carelessness in terminology is painful to the well-tempered anatomist's ear, like carelessness in performing music. Anatomists are sufficiently conversant with their craft to notice, and correct, slapdash usage of anatomical terms. I advise you not to take this lightly: they *will* correct you. Your professors will let you know that bad anatomical grammar is something up with which they will not put. And when you write an essay in your class, or an article for a journal using the phrase "lying on your stomach" (a quite impossible stunt, if you think about it), it will be [sic]ced, and [sic]ced good and hard. You mean "lying prone," and it should be written that way. Don't be defensive about this; it's not a sign of arrogance when your invigilators are vigilant. That's what they are there for.

"Whether you softly speak
Or whether you loudly call
Distinctly! Distinctly speak
Or do not speak at all."
Unknown

Solecisms in Gross Anatomy

A common mistake that anatomy students make is in their use of the terms "arm" and "leg." These are perfectly legitimate terms to use in anatomy, but be sure you use them correctly. Lay people refer to the entire upper extremity as "the arm." For an anatomist, "the arm" is more specific: it is the part of the upper extremity between the shoulder and the elbow. From elbow to wrist is the *forearm*. Similarly, the "leg" is *only* from the knee to the ankle. Between the torso and the knee is the "thigh."

You should approach learning anatomy like you would approach learning a new language. If you learn what the words mean, you'll have gone a long way toward understanding the anatomy. And you should be aware of the customs of the discipline. Most anatomists are admirers of those traditions. One tradition is this: the best way to learn what a word means is to break the word into sections, and learn the definition of each section.

Here's a simple example: "Endocarditis" is a combination of several parts, a *portmanteau word**, if you like. The "endo" part comes from "endon" which means "within." "Cardi" comes from the Greek word "cardia" which means "heart." "Itis" means "inflammation." Can you see how the word "endocarditis" refers to inflammation of the inner lining of the heart?

These helpful derivations of the various parts of medical terminology are readily available in any medical dictionary. I use *Stedman's Medical Dictionary*, the full unabridged version, and it is excellent. In some cases, your plain English dictionary will help; I use *New Oxford American Dictionary, 3rd edition*, because it looks so impressive on its podium here in my study. Your high school Latin teacher told you that you'd need to know Latin in your advanced education, and as it turns out, your Latin teacher was right. You can learn most of the words you missed or neglected in her excellent high school Latin course, with a *Stedman's Medical Dictionary*. Mine is the 27th edition.

**Portmanteau*: a word that combines the meanings of two others. This designation comes from a line in Lewis Carroll's *Through the Looking Glass*: "You see it's like a portmanteau... there are two meanings packed up into one word."[5]

Let's take another example. The lateral antebrachial cutaneous nerve. That's a mouthful, and if I were to simply learn to say the name by memorizing "nonsense syllables" by rote, learning the name would be unnecessarily difficult. Instead, I could make myself aware that "lateral" means "away from the middle of the body." *Webster's New World Dictionary* tells me that the prefix "ante" means "in front of." In front of what? The brachium. *Webster's* tells me that this word comes from the same Latin word ("brachium") that the word "embrace" comes from, and it refers to part of the upper extremity between the shoulder and elbow (the arm). "Cutaneous" comes from the same word as "cuticle," meaning the outer layer of the skin. So I can determine by its name that the lateral antebrachial cutaneous nerve is located on the outer side (lateral) of the front (ante) of the arm (brachium), and it serves the skin (cutaneous).

You'll want to work hard to be accurate in your speech. A little time spent establishing a friendship with your dictionary and medical dictionary will pay big dividends in learning Anatomy. Don't just glance at the definition; *help* yourself by checking out the derivation of the word. I promise you, you'll be glad you did. Remember: *chance favors the prepared mind.* If you prepare your mind with Greek and Latin derivations of the words you've encountered in your study of Anatomy, chances are you'll be able to figure out new anatomical words as you encounter them. And the meanings of the words will be far more memorable, learned this way. You will have a "leg up" (figuratively speaking) on those who learn the words by rote.

Branches and Tributaries

Arteries have branches; veins have tributaries. These are rules that are often not followed in conversation in the Anatomy Laboratory, and in other places. When an artery takes blood from the descending aorta, that artery is a *branch* of the descending aorta. When a vein drains blood into the inferior vena cava, that vein is a *tributary* to the inferior vena cava. In turn, the inferior vena cava is a *tributary* to subclavian vein. Pedantic, you say? Maybe so. Nonetheless, most anatomists consider this an important distinction.

Abbreviations Common in Anatomy Textbooks

cf: means "refer to" or "cross-reference." It means, "You should also look at this other reference, for clarification or further information concerning the subject you're reading about."

ff: "and the pages that follow."

Therefore, if the narrative says "Choking. cf: Notespak this chapter, page 15 ff," that means: "Choking. Refer also to Notespak this chapter, page 15 and the pages that follow."

N.B.: The abbreviation "N.B.:" comes from the Latin words *nota bene*, which mean *note well.* When this abbreviation appears in the narrative, it indicates that what follows is important.

Let's try that on the following topic.

Assume the (Anatomical) Position

N.B.: You must become aware that anatomists *assume the anatomical position*, in their descriptions. *The anatomical position means standing erect with feet together and palms forward.* It's not complicated, but it's important.

Let's say your anatomy professor asks you: "Is the thumb located *lateral* to the pinky, or *medial* to the pinky?"

You may answer, reasonably, "It depends on which way you hold your hand." *Wrong!* The professor, like all anatomists, assumes the anatomical position. When you are standing with your feet together and palms forward, the thumb is, of course, lateral to the pinky. From an anatomist's point of view, the thumb is *always* lateral to the pinky.

Remember: the anatomical position is *assumed*, when describing anatomical relationships. In common anatomical parlance, one does not say, "In the anatomical position, the thumb is located lateral to the pinky." No. An anatomist says, "The thumb is located lateral to the pinky."

Stedman's Medical Dictionary defines the anatomical position thusly: "the erect position of the body with the face directed forward, the arms at the side and the palms of the hands directed forward; the terms posterior, anterior, lateral medial etc., are applied to the parts as they stand related to each other and the axis of the body when in this position."[6]

O.K., anatomical position means you stand thus and so. With your feet together and your palms facing forward. Pretty easy. Now let's take it one step further.

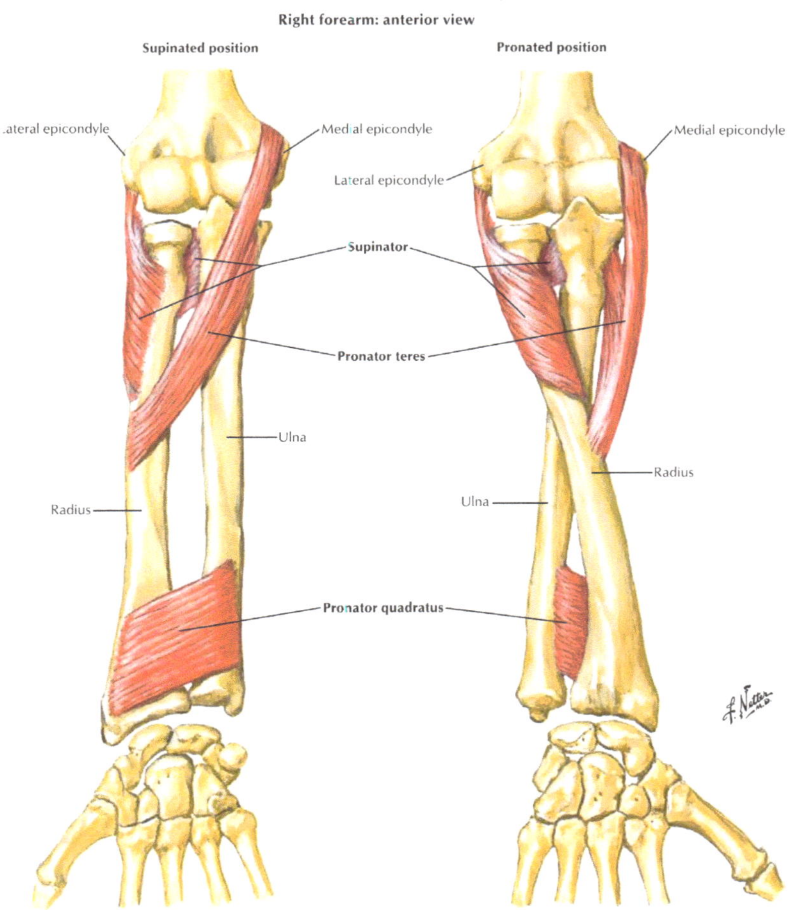

Netter Atlas of Human Anatomy, 5th Edition 428.
Copyright 2011 Elsevier Inc All rights reserved.
www.netterimages.com 4388

Have a look at *Netter 5th ed*, 428, entitled "Individual Muscles of Forearm: Rotators of Radius." We are looking at the forearm from anteriorly (from the front). In the drawing on your left, the person is in anatomical position, with the palm of the hand facing us, and you can see that the radius is located lateral to the ulna. In fact, when your anatomy professor speaks of the forearm, he or she will assume that you know that "radial" equals "lateral," and "ulnar" equals "medial." Why? Because we always assume the anatomical position.

"Dorsal" means toward your back, and "ventral" means facing the same way as your belly faces. In other words, dorsal faces the same direction that the back faces, and ventral faces the same direction as the chest faces. Hey! That's easy!

Not so fast. What can be confusing is the way anatomists use the words "dorsal" and "ventral" to refer to your feet, and your tongue, where "front" and "back" don't seem quite so applicable. How can the *top* of your tongue be considered the *dorsal side?* How can the *top* of your foot be considered the *dorsal side?* The top of your tongue does not face the same direction as your back faces.

N.B.: The *best way to fix in your mind how "dorsal" and "ventral" apply to all structures is to think of your four-legged friends.* The dorsum, or dorsal surface, of your dog's tongue is indeed on the side corresponding to his back. The ventral surface of his tongue faces the same way as his chest faces. What about your dog's feet? The top of his foot faces the same direction as his back faces, and therefore it is the dorsal surface of his foot. And so is yours.

Have you ever heard of plantar warts? They occur on the bottom of the foot, and their name is *not* derived from the idea that they appear on someone doing some planting. Those warts are so named because the bottom of the foot (the "sole" of the foot") is called its "plantar surface." Plantar flexion, therefore, flexes the foot in the direction of the bottom of the foot, and its opposite is dorsiflexion.

If you have a bad back like I do, you don't pick up your underwear from the floor by bending over; you hook your great toe (your hallux) under it, and flip it up to catch it. In doing so, you go from plantar flexion in sliding your toes under your drawers, to dorsiflexion when you kick those tidy-whities up in the air.

Flexion and Extension

For the anatomist, "flexion" and "extension" are terms that people toss around (like the dirty underwear) as if their meanings were obvious. In most cases, they are. Here is the general rule: flexion means decreasing (closing) the angle between two body parts. Extension means increasing (opening) the angle between two body parts. If I flex my forearm at the elbow, what I'm doing is decreasing the elbow angle, and my wrist approaches my shoulder. In most cases, it's just obvious.

Here's an example: if I'm standing upright and I take a bow, I'm changing the angle at my waist from 180 degrees to about 90 degrees. I have flexed my body by bowing. I extend my body by standing upright again.

If I'm showing off my strength by demonstrating my biceps brachii muscle, I'm flexing my forearm at the elbow. Extending my forearm at the elbow is straightening out my entire upper extremity.

Pronation and Supination

The words "pronation" and "supination" refer to the relationship between radius and ulna. Notice *Netter 5th ed.* 428, "Individual Muscles of Forearm: Rotators of Radius," on page 31. This plate demonstrates the pronated position and the supinated position of the forearm. *Here's the universal way to remember pronation and supination: if your hand is in a position to hold a bowl of soup, you're supinated ("soup-inated").* If you remember that much, you'll know that the opposite of that, with your hand turned over and dumping the soup out, is the pronated position. These terms can also be applied to the foot, relating to the inward roll of the foot (pronation) and outward roll of the foot (supination) during normal walking.

N.B.: To demonstrate pronation and supination correctly, the forearm needs to be flexed at the elbow. Here's why: allow your upper extremity to hang loosely at your side. Now rotate your hand back and forth in the manner shown in the drawing above, *Netter 5th ed.* 428. What moved? Typically, most of the movement occurred at the shoulder. The radius and the ulna did not change much in their relationship one to the other. If we want *that* change in relationship, we want the movement to occur at the elbow joint. Flexing the forearm at the elbow accomplishes this for us.

So flex your forearm at the elbow, move your hand from pronation to supination and back again, and notice the radius crossing over the ulna. There! That's more like it.

You Should Know Right from Left

In Anatomy, the terms "right" and "left" almost always refer to the patient's (or cadaver's) right or left. An instructor pointing out something on a PowerPoint slide may say, "on the right side of the picture," or "towards the left side of the drawing," or even "the structure on your left," but if the instructor says, "the left ear," it refers to the *patient's* left ear. The anatomist teaching your class is speaking as a doctor would.

Pronunciation

There's an old story about a husband and wife traveling through Damariscotta, Maine arguing about the correct pronunciation of the name of the town. He insisted on placing the emphasis on the second syllable, while she placed it on the fourth syllable. Finally, to resolve the argument, they stopped at a fast-food restaurant, and politely approached the clerk.

"Would you please pronounce the name of this place, carefully and correctly?" To which the clerk articulated, slowly, "O.K., the name is "Mc - don - ulds ."

Correct pronunciation is important, and can be found in any medical dictionary. You should pronounce "diverticulitis" with emphases on the first and fifth syllables.

The Eponym

Stedman's Dictionary defines "eponym" as "The name of a disease, structure, operation or procedure, usually derived from the name of the person who discovered or described it first, or made the structure famous. [G. *eponymos:* named after]"[7] "Bell's palsy" is a common example. And "Achilles tendon."

Anatomical nomenclature is an ancient language. It's euphonious. There is some movement afoot (no pun intended) in Anatomy to dispose of eponyms. The part of the eyeball where the aqueous humor drains (preventing glaucoma) is called the canal of Schlemm, but modern anatomists are working to change its name to sinus venosus of the sclera. Which is more euphonic?

Anatomical nomenclature depends on the classic languages, Latin and Greek, and has a keen eye for description. The names were added to the discipline, little by little, over the millennia. Along the way, sometimes the structures were named descriptively, and at other times credit was given for conspicuous research in one organ or another. The hierarchy in anatomical nomenclature has occasionally rewarded such a person by attaching his or her name to the structure. One commonly known eponym is Bartolommeo E. Eustachio, for whom the eustachian tube is named. It has been an excellent way to honor a lifetime of ground-breaking work in a specific field.

Notice my words: it *has been* excellent in that way. Things are changing nowadays, and the trend is to move away from eponyms, and to saddle each structure with a name more descriptive, and more dependent on the Greek and Latin. For example, the name of Dr. Eustachio's tube is being changed to the pharyngotympanic tube. The name indicates that the tube goes from the pharynx to the tympanic cavity. The referent is the same, but the proper name for it is changing. Be prepared, though: when they refer to the pharyngotympanic tube, your patients will use the eponym.

There is a rich tradition of such nomenclature, and when we switch our words, we alter reality a bit – to a better reality, in the eyes of the reformers. Ordinary and traditional usage and dictionary definitions are no longer binding with the progressives – they perceive a moral imperative to progress – and the result is that countless researchers and teachers whose work has enhanced our understanding of ourselves will be neglected, while the art of Human Anatomy is, in the eyes of your humble correspondent, bowdlerized.

Some of us bristle at this schoolmarmish revisionism because we feel it unjust to ignore those revered pioneers in the discipline we hold dear. In the name of progress, or political correctness, we risk expurgating the elegant work done by our forefathers in Gross Anatomy. The iconoclastic revisionists are out to change an ancient language, and for what? Dr. Rene Le Fort was a French surgeon and gynecologist. In 1901 Dr. Le Fort published a treatise called *Étude expérimentale sur les fractures de la mâchoire supérieure* involving his experiments with maxillary fractures. To perform these experiments, Le Fort used intact cadaver heads, and delivered blunt forces of varying degrees of magnitude, from various directions. From these tests, he determined that predictable patterns of fractures are the result of certain types of injuries, and concluded that there are three predominant types of mid-face fractures when your face hits the steering wheel. (Wear your seat-belt!) These classifications, Le Fort I, Le Fort II and Le Fort III, have served the practice of surgery to the face for over 100 years. Are we now supposed to stop honoring Dr. Le Fort's groundbreaking research and call the three fractures horizontal fracture, pyramidal fracture and transverse fracture?

When it comes to getting rid of eponyms, I'm a Tory (not "tori," which are bony bumps). I'm like those who remained loyal to the British crown. By my lights, it's more than courteous, it's justice, to honor those men and women who've done such important work. Through a lifetime of service to the discipline we hold dear, they've made the way a little better for us. To me, it's like when I use my computer, the powers that be have decided the default font will be Calibri. Was there ever a worse font than Calibri? Times New Roman is easier to read, and much more pulchritudinous, as well.

In some of the drawings in *Netter's Atlas*, you'll see the name *and the eponym*, together on labels. There are two examples on one plate: *Netter 5th ed.* 81, entitled "Eyelids." See the middle drawing labels "Superior tarsal (Muller's) muscle (smooth)," and, farther down the page, "Tarsal (meibomian) glands." Along with the neologisms, I'd advise you also to pay attention to the traditional names of things. "Make new friends, but keep the old; one is silver and the other's gold" is still good advice, because in the literature, "Stenson's duct" will appear as frequently as "parotid duct," and you'll need to know what that is. And don't forget your patients; they won't be calling their drooping cheek "peripheral facial paralysis"; they'll just say "Bell's palsy." And they should.

Part of the anti-eponym movement in Anatomy is the rumor that in many cases, the person's name was applied, *not* because that person did ground-breaking work on that particular structure. Rather it was given by some society or group of influential colleagues, to honor a person whose work *may be* unrelated to the structure. In other words, the person honored may or may not have done significant work on that structure. This opinion, along with a certain tendency to condescend toward the past, has fueled some of the growing sentiment against eponyms. Nonetheless, I believe there are enough of us principled types to avoid changing the name of Colles fracture, for instance, to some more descriptive, less eponymic, name.

You might ask, "Why, in the paragraph above, is Muller's name capitalized, and Meibonius'* name not capitalized? The rule is this: when the word is *possessive* (as in "Muller's" above), it should be capitalized. When it is merely an *adjective* (as in "meibomian") it is not. Thus the term "eustachian tube**" is not capitalized (*Netter 5th ed.* 63, entitled "Pharynx: Median Section"). If it were called "tube of Eustachius," the person's name would be capitalized. Be aware, however, that these rules are not always strictly adhered to.

> *Dr. Meibomius, himself, had two names. My *Stedman's Medical Dictionary, 27th ed.* lists him thusly: Meibom (Meibomius), Hendrick (Heinrich). *Stedman's* goes on to say "German anatomist, 1638-1700."[8]
> **Parenthetically, the tube that bears Eustachius' name has three names: auditory tube, pharyngotympanic tube, and eustachian tube.

The New Oxford American Dictionary, 3rd edition defines the word "antonomasia" in the following way: "the use of a proper name to express a general idea, e.g. a Scrooge for a miser."[9] It comes from Greek: "anti" = "against," "onoma" = "a name." Antonomasia and the use of eponyms are similar concepts.

Several Names for Every Structure, and Vice Versa

Anatomical language is an esthetic tool as well as a scientific one. Like other resplendent languages, the language I call "anatomania" often has two or three names for a particular structure. The chin has three words that refer to it: mental, genial, and chin. The duct from the parotid gland to the oral cavity is Stenson's duct, and it is also parotid duct. Both names are accurate, equally correct. And let's consider the common peroneal nerve: the gods of Anatomy, who cannot permit the freezing of anatomical nomenclature in its tracks, are gradually changing its name to "common fibular nerve." Dr. Netter, in his *5th edition*, sometimes accommodates this trend by putting one name in parentheses. For example, on *Netter 5th ed.* 483, entitled "Muscles of Hip and Thigh: Posterior Views," the nerve is labeled: "Common fibular (peroneal) nerve."

Anatomical *structures* almost always have more than one *name.* Conversely, there are, in many cases, *names* that refer to more than one *structure.* The word "sinus" refers to air-filled spaces, blood-filled spaces, and lymph-filled ones. It means different things in different contexts. Here's a partial list:

Ten Different Kinds of Sinuses

1. Carotid sinus – contains oxygenated blood. *Netter 5th ed.* 124, 129
2. Dural sinus – contains deoxygenated blood. *Netter 5th ed.* 102, 103
3. Paranasal sinus – contains air. *Netter 5th ed.* 48, 49
4. Pericardial sinus – is a potential space between heart and posterior aspect of pericardial sac. *Netter 5th ed.* 209
5. Aortic sinus (of Valsalva) – is a space immediately above each cusp of the aortic valve.[10] *Netter 5th ed.* 217
6. Coronary sinus – contains deoxygenated blood. *Netter 5th ed.* 215
7. Lactiferous sinus – sometimes contains breast milk. *Netter 5th ed.* 176
8. Sinus venarum cavarum – contains deoxygenated blood from SVC and IVC in the right atrium. It is also called "sinus of the vena cava." Not identified in the Atlas, it is located right where Dr. Netter has labeled "interatrial septum" on *Netter 5th ed.* 214, top.
9. Renal sinus – contains fat. *Netter 5th ed.* 311

10. Sinus of Morgagni – contains nuthin'. It's the ventricle of the larynx (*Stedman's 27th ed.* 1644), shown but not labeled on *Netters 5th ed.* 63. Because of the presence of the ventricle, the false vocal cords are labeled "ventricular folds" on *Netter 5th ed.* 78.

This "different names for one structure" and "different structures with the same name" is something you'll need to get used to. In Lewis Carroll's *Through the Looking Glass*, written in 1872, there's a discussion that reminded me of the plight of students learning nomenclature in Gross Human Anatomy.

"But 'glory' doesn't mean 'a nice knock-down argument,'" Alice objected.

"When I use a word," Humpty Dumpty said, in rather a scornful tone, "it means just what I choose it to mean—neither more nor less."

"The question is," said Alice, "whether you can make words mean so many different things." In Human Anatomy, Alice, we can.

Speaking of Blood Vessels

The opposite of drainage (veins) is perfusion (arteries).

Claudication is blood flow obstruction.

Arteries have branches. Veins have tributaries.

Let's think for a moment about facial artery. It branches at about the bare area of the mandible. Now, let's say you want to refer to the place where submental artery branches from facial artery. Should you call that "the submental branch of facial artery," or simply "submental artery?" Either is fine, and equally valid.

Ellipsis

Early in your Gross Anatomy studies, you will notice something peculiar in the way anatomists talk, both in the laboratory and in a lecture. This is the use of ellipsis: the intentional omission of a word or words that normally would appear. A common example is someone saying "if possible" instead of "if it is possible." *Stedman's Medical Dictionary* defines ellipsis this way: "Omission of words or ideas, leaving the whole to be completed by the reader or listener. [G. *ek,*- out, *leipsis,* leaving.]"[11]

Whereas most people would say "This is the vagus nerve coursing around the aorta," an anatomist might say it this way: "This is vagus, coursing around aorta."

You will see plenty of ellipses in your study of Anatomy. In fact, in many circles it is considered better form to say "Here is internal jugular vein, coursing through carotid sheath," rather than "Here is *the* internal jugular vein, coursing through *the* carotid sheath." A certain amount of ellipsis is the vernacular of Anatomy.

The Importance of Punctuation

When referring to the Anatomy Laboratory, be careful with your punctuation. A colleague's wife once told her friend, "Those dead bodies in the Anatomy Lab are my ex-husband's."

Cadaver Lab Lingo

There are terms and phrases that are used loosely in the Anatomy Lab, and you will very soon get accustomed to using them yourself. One example is when a professor says a structure is "heading right for the tabletop." That is referring to the top of the cadaver table, of course. A cadaver in the supine position will display musculature in the lower extremity and "perforating branches" that course almost directly *dorsally*, or in the direction of the back of the cadaver's thigh. In other words, they course in the direction toward the top of the table that lies under the cadaver. We therefore say that they "head right for the tabletop."

I would caution you: we don't use the phrase "flipping the body" to describe turning the cadaver over. That terminology comes pretty easily, but it seems disrespectful. Furthermore, it's not actually what we do. What we do in changing a donor body from supine position to a prone position is a deliberate, multi-person effort that's described in detail in a section entitled "Turning the Body" in Chapter 9 of this book. We like to hear people refer to that exercise as "turning the body." Thank you; that's much better.

Contralateral, Ipsilateral, Homolateral

Contralateral: belonging to or occurring on the opposite side of the body
Ipsilateral: belonging to or occurring on the same side of the body
Homolateral: same as ipsilateral

Etymology

You will remember the terminology better if you know the derivation of words. Sometimes the derivations are so audacious that they make the word easy to remember. Check out the mammillary bodies on the inferior surface of the hypothalamus, as they're drawn in your *Netter Atlas 5th ed.* on page 106, entitled, "Cerebrum, Inferior View." Their name makes you think of "mammary," doesn't it? Look up the derivation of the word "mammary" in your *Stedman's*, and you will read: "relating to the breasts."[12] The word is a cousin to the word "mammillary," which refers specifically to the nipples. When you make this association, you won't soon forget what those little bumps are named. (Actually, they *are* shaped like tiny nipples.)

When you come to a new word, break it down. For example,

Macrocephaly: composed of:
macro- = large, or mega-
cephalic = head

Which of the following is a synonym for *macrocephalic*? (*Stedman's 27th ed.* 1051)?

nonmicrocephalic
*megacephalic
bigguscephalic

And its antonym is *microcephalic.*

Some Interesting Anatomical Terms

Splanchnic: when you get to the splanchnic nerves, you may wonder why they are called that difficult-to-spell name. Turns out my *Stedman's 27th ed.* 1675 tells me that "splanchno-" comes from the Greek word for "viscus," which is the singular form of the word "viscera." The splanchnic nerves run to the viscera (the "innards"). Splanchnology is the branch of medical science dealing with the viscera.

Crus: you'll find the word "crus" frequently in Anatomy. It looks almost like "cross," and that's what it means. Your *cruciate ligaments* cross over each other, leading to their name.

Popliteal: the popliteal fossa is the diamond-shaped space located on the posterior side of the knee joint. The word "popliteal" comes from "poples," which is derived from the Latin, meaning "the hollow of the knee"[13]

Acro-: the prefix "acro-" means the highest point, the topmost. Therefore, would it surprise you that the city of Akron is built on a hill? I want you to know that the prefix for "shoulder" is "omo" in the same way that the word for "kidney" is "renal." Now, where do you think the "acromion" is located? It's the topmost bony point on your shoulder. "Acropolis" has the same derivation: highest point in the "polis," or city.

Meninx: many of us are familiar with the word "meninges," referring to the tough fibrous membranes that surround the structures of the central nervous system. "Meninges" is a fairly common word, but it's plural. What is the word that refers to just one membrane? Meninx.

That brings me to one of my favorites:

Pachy-: comes from a Greek word for "thick," and is pronounced "packy." In fact, "pachymeninx" refers to the dura mater: thick meninx. It is to be distinguished from the leptomeninx, which refers to the arachnoid mater and pia mater, combined. ("Lepto-" = light, thin and frail.) Parenthetically, the thick-skinned elephant is a "pachyderm."

In my *Stedman's Medical Dictionary*,[14] there is almost a full page of "pachy-s," including pachyglossia, a thickening of the tongue, and my favorite: pachydermoperiosteosis. Let's break that word down: a thickening ("pachy-") of skin (derma = skin) together with a type of new bone formation (periostosis: "peri-" = around, "osteo-" = bone, and "osis-" = production or increase, physiologic or pathologic).

Visceral and parietal: If I were a human organ like the liver, parietal peritoneum is the paint on the walls of the room I'm in. Visceral peritoneum is my shirt. The air in the room (between the walls and my shirt) is in the peritoneal cavity.

The calvaria is the upper dome-like part of the skull that encloses the brain. Calvaria, not calvarium! What is the "calvarium?" Mark this: *"calvarium" is not a word.* The word "calvaria" is singular; the plural is "calvariae." The word "calvarium" is always incorrect. After the calvaria, the rest of the skull is the "facial skeleton."

Skull vs cranium: What is the difference between the words "skull" and "cranium?" *The skull includes the mandible, whereas the cranium is the skull without the mandible.* These particular definitions, however, are not always strictly observed.

Genotype is a person's genetic makeup.

Phenotype is what you see. The observable characteristics of an individual, determined by the genotype and the environment.

Tubercle is just a bump. Tuberosity is a large bump.

Chirurgery: "chirurgery" means "surgery"[15].

Pelvis: derived from the word "basin" (L)

Synarthrosis: joints that don't move. Memory aid: some people tend to think certain dance movements are sinful. Demonstrate synarthrosis by trying not to move in ways that are sinful. (I've never been to a joint where they move like that.) An example of a synarthrosis joint is gomphosis: the joint between a tooth and the bone that encases it.

Mastoid process: The stem "mast-" = "breast." The suffix "-oid" means "resembling." Early anatomists were almost exclusively male, and they thought this bony process resembled a breast. Have a look at the mastoid process (*Netter 5th ed.* 15), consider its similarity to a breast, and you'll agree: clearly, those early anatomists needed to get out more.

Pterygo-: part of a word you'll see often in your study of the head & neck. It comes from the same root as the word pterydactyl does, which refers to a winged pre-historic animal. *Pteryx*, or *pterygos* is the Greek word for "wing."

Petrous: comes from the same root as "peter," meaning "the rock." The petrous portion of the temporal bone is the hardest bone in the body.

Conchae: (pronounced "CON ky"): shaped like one of those shells you pick up on the beach that is called a conch. Early anatomists thought part of your ear resembles conchae, and named it accordingly. And those scroll-shaped bones on the walls of your nasal cavity are chonchae, as well. Physicians call them "turbinate bones."

Hyoid: the word "hyoid" means "U-shaped." Given this information, the word "hyoid" could be used to describe the *shape of* the mandible, or the shape of the thyroid cartilage, or even that of the tracheal rings (*Netter 5th ed.* 15, entitled "Bony Framework of Head and Neck"). They are all open in the back. The word could also be used to describe those awkward smocks, tied with strings, that they make you wear in the examining room of the Emergency Department; that smock *certainly* is open in the back. Specifically, the word "hyoid," alone, refers to that bone in your cervicomental angle. What is the only airway structure that is *not* open in the back? Cricoid cartilage: it is a complete circle, and it surrounds the airway.

Sella turcica: literally translated from Latin: a Turkish saddle. It's a bony structure in the skull, and it's actually shaped like a saddle.

Articular eminence: a bump (eminence) where two bones come together in a joint (articulate with each other).

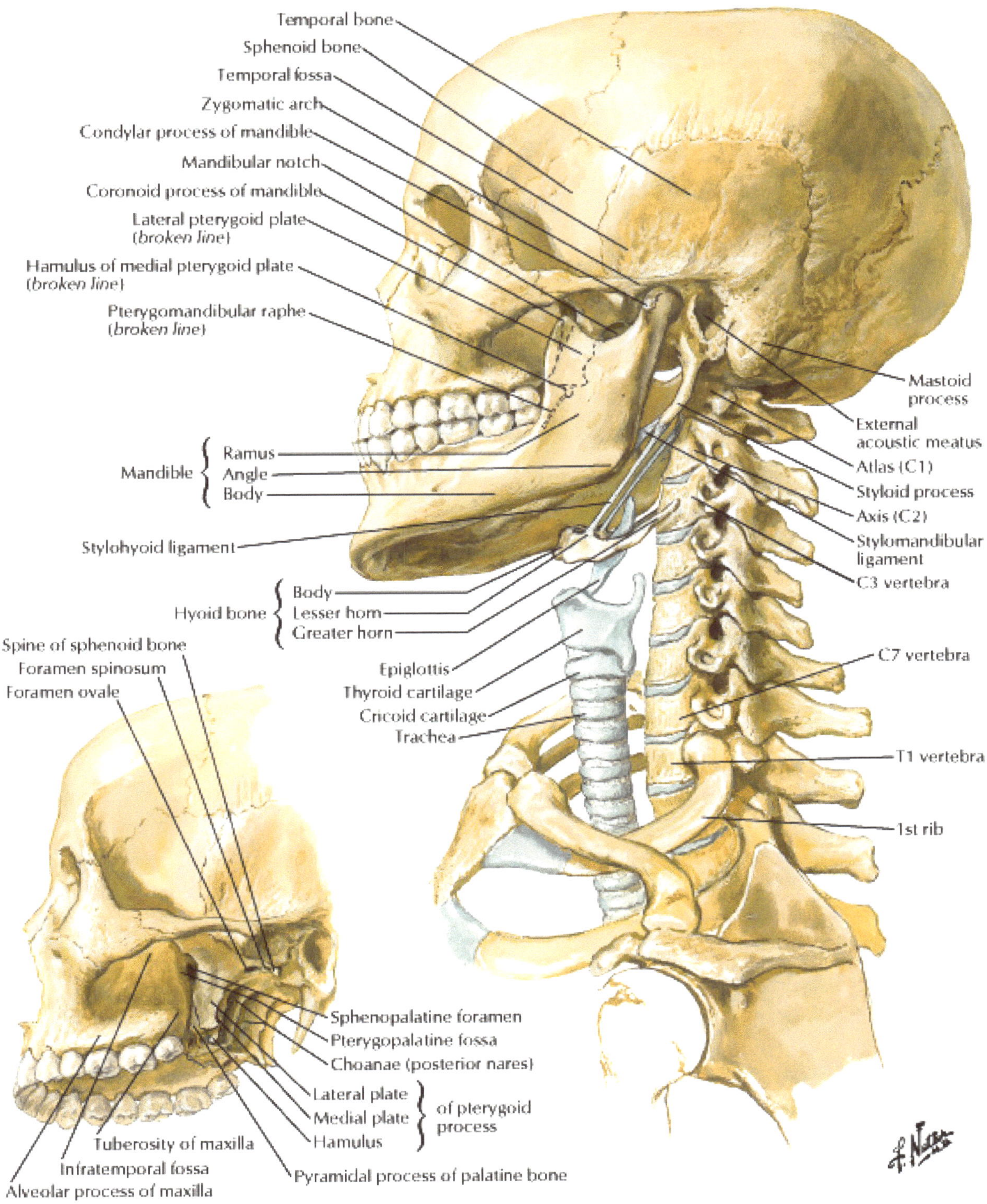

Netter Atlas of Human Anatomy, 5th Edition 15.

Using the Language of Anatomy

Practice using the language of Anatomy. Don't say "leg" when you mean "lower extremity." Don't say "calvarium" because there's no such word. Say "calvaria" for a single skullcap, and "calvariae" if there are several. Be aware that your food doesn't go directly from your mouth to your esophagus; it goes through the pharynx on the way. Don't say "pain in my belly," say "pain in the lower right quadrant of my abdomen," and then get to the hospital for your appendicitis. Be precise in your speaking.

When I first joined the ranks of the anatomists, I noticed something that you will notice, too. Anatomists are so immersed in their discipline that they don't realize that you aren't. They make statements that are so obvious to them that they assume they're obvious to everyone. And you won't study Anatomy very long before you'll be doing the same thing to your friends. You'll see someone "get the wind knocked out of them," and you'll remark about a stunned phrenic nerve, or diaphragm spasms, concepts that seem obvious to you. To your companion, it's Greek.

If you're studying one of the health professions, you will soon realize that *anatomy is the language of medicine.* Metaphors are abundant: as I said before, anatomy is the floor of many rooms, like physiology, embryology and pathology. Anatomy is the fabric of the practice of healthcare. It is the foundation, the *sine qua non* in medicine, with all its various manifestations, and a thorough knowledge of anatomy is absolutely essential to the practice of caring for other people.

CHAPTER 3A:
Discourse with the Donor

I had been asleep for six hours, and morning was approaching. I heard a voice in my ear, and I smelled formaldehyde. When I reluctantly opened my eyes, there she was again, looking at me from the foot of the bed: Specimen 2436, from Table 6.

Last night she was particularly loquacious. "Those miscreants at Table 8 opened their donor's chest by making a "Y" incision like a coroner makes. That's an autopsy incision. Incisions in the anatomy lab are supposed to be horizontal, from acromion across the clavicles to the other acromion. like swinging doors in the old cowboy movies. These kids don't follow directions. The correct incision is *directly athwart* the thorax." She clearly was becoming agitated at the thought.

"You are familiar with a coroner's incision? You yourself don't have an autopsy incision. Were you..." I was being very careful, here, "...unattended?"

"No. It was not an unattended death, so I didn't require autopsy. I died from the inside out, like a tree dies. It took a while..." She seemed to turn that over in her mind for a few seconds, and then said, "Ironically, dying was the last thing my body wanted to do."

"And indeed it was the last thing your body did. And then your body ended up in my laboratory," I said. "Were you alone, at the time?"

"Sadly, my family was all in attendance, when I escaped this mortal coil." She smiled darkly as she reflected. "The doctor gave me pain medication and told me I'd be taking it for the rest of my life. Then I looked at the bottle, and it said, 'Good for three refills.' Three refills is about how long I lasted." She looked at me to confirm that I understood her macabre little joke. She was smiling, until she saw the sorrowful way I was looking at her.

She said, "In the end, there was no Grim Reaper, a gaunt figure in a hooded cloak, wielding a deadly scythe, separating my soul from my body with one lethal swipe. Nothing like that." She paused, reflecting for a moment. "And there were no heroics; no emergency surgery, and ultimately no autopsy. The reason I'm conversant with autopsies is that in life I was an English professor, and I have read thousands of essays on hundreds of topics. You know how it is: an outstanding performance by a student sticks in your consciousness forever. One such memorable essay was about autopsies."

"You were an English professor? I hadn't the slightest idea." I began to form in my mind an image of this sinewy person standing in an auditorium, lecturing. It was an endearing thought.

"Oh, yeah. And you've caused me to cringe many times in the lab, as I silently corrected your grammar."

"Seriously? Me? Bad grammar? Such as?"

"Such as today you told the students at my table they needed to carry their discussion a little farther. The word is "further." You said carry a discussion farther! *Seriously?* And *then* you said, 'Before you leave the lab, show your dissections to Dr. Midias and I.' It's *me*, not *I*!"

"I said that?" I said, quietly. "Good grief."

"Yes. And you said we cadavers should be kept 'partially nude." Professor, "nude" means just that: nude. You're either nude or you're not. You should have said, 'partially clad.'"

"I never thought about those…"

"Bro, *that's your job.* Sometimes your solecisms and malapropisms sound… bibulous! These kids are impressionable; you can't go around talking like English is your second language. You're careful about what you say; you should be careful about how you say it."

I was completely caught off guard by this. Frantically, I searched my mind for a riposte.

"You're surprised that I was an academic, like you," she continued. "You just think of me as an old harridan," she said. "No. Quite the contrary. I'm a word person. That's why I get offended when students use words like 'stiffs,' and "corpse" when referring to those of us on the tables. They use words like 'road kill,' and some other pretty disgusting dysphemisms. They should have more respect."

"I'm not sure that those terms constitute dysphemisms. More like sobriquets. I heard a student today refer to his "guns." Do you think he should have called them biceps brachii muscles? I think some of these are simply nicknames. Harmless."

She whirled around. "Oh, yeah? What about calling me 'knock-kneed'? Or using the word 'caboose' for *this!*" She pointed vigorously to her beautifully dissected gluteus maximus, gluteus medius and gluteus minimus. They were separated perfectly, and the layers were stacked. I couldn't help noticing that sciatic nerve was artfully preserved, as it erupted from under piriformis and draped over her lateral rotators.

But 2436 was clearly upset. I made a mental note to address this in my next lecture, but I didn't mention it to her, because, frankly, I was afraid of what she might instruct me to say.

So I said, "I think you're making 'way too much of this."

"Everybody seems to have little monikers for my various parts, and they're not always kind. Young people nowadays! Coarseness is their lingua franca. And, I have to say, *mutatis mutandis*, for the lab instructors. All that twaddle is unbecoming, in a serious venue like the anatomy laboratory. It's important work we're all doing, and the students should have more respect, in my view."

I was a little taken aback by her allegiance to the lab. She went on. "It's as if those instructors suffer from gymnophobia, or something. They're like Doc Martin, the surgeon who can't stand the sight of blood."

"What's gymnophobia?" Just as the question was coming out of my mouth, I sussed that I should have known the answer already.

She looked at me incredulously and cocked her head to one side, like a dog watching TV. "Gymnophobia, *Doctor*, is the morbid dread of the sight of a naked person. It even means dread of the sight of an uncovered *part* of the body. And Lord knows we have plenty of *those*, lying around the lab."

"I never knew that word before…"

"It might interest you to know that the word 'gymnasium' literally translated from the Greek means 'school for naked exercise.' Thankfully, gymnastics is not practiced that way nowadays. But gymnophobia abounds." She smiled demurely. "Especially among Presbyterians" she said, "like you."

"That reminds me," I said. "Isn't Elizabeth at your table? I'm afraid she's got a touch of… *gymnophobia*, and the result is she's failing the course. I'm concerned about students like her, who just aren't doing the work. They're not making the effort; it's apparent in their test scores. She needs to put in more time, stay after class, come and see me in my study. She's not doing any of the things that will allow her to be successful." I felt like I was lecturing my teenaged daughter. "If she doesn't mend her ways, our sweet Elizabeth just may flunk out."

She leaned toward me, raised one eyebrow, and said softly, "Habibi, paying attention is not Elizabeth's problem. She's got issues you've never even thought about."

"Really?" I hardly knew the diffident young lady at Table 6. "How so?"

"When Elizabeth goes home every night from medical school, she faces a family with high expectations regarding her domestic duties. They expect her to cook their meals, do the grocery shopping, and attend her little sister's school events. They simply don't understand the level of commitment it takes to be successful at a professional school."

She stopped for a minute, steaming. If she'd had a blood pressure, it would have been soaring. But, alas, she did not. "I know," she said, "because I've heard my students talking to one another as they dissect. This week Elizabeth's aunt is coming for a visit, and Elizabeth is not only cooking the meals, she's giving up her bedroom, where she studies, to her aunt."

I was smitten by the fact that she referred to the students at her table as "my students." For a moment, I regarded what she had said. "Who knew all that was going on with Elizabeth?" I was filled with remorse. "Maybe she and I should have a chat."

"Each of the hundred and twenty-five students has problems. There are lots of these sorts of things. Some even have mental or physical problems."

"Really? Like what?"

"Well… hysteria, panic attacks, clinical depression..."

"There's a counseling center for this," I interrupted. "They should go there. They can get accommodations just by asking."

She continued, ignoring me. "…bipolar disorder, fainting, flush…"

I interrupted her again. "The problem is that some students will game the system and claim a disability, in order to get accommodations that give them an unfair advantage." It was a weak, ineffective reply, and we both knew it.

She abruptly stopped her litany of disorders. "Doctor, I want to let you in on something you should know: *everybody in your life, including each of your students, is fighting a battle that you know nothing about. You need to consider things from their point of view. Some of them have peculiar circumstances that would shock you.*" She put her partially-dissected hand firmly on my shoulder and looked at me with the intensity of a general sending a soldier into battle. "Your job, my friend, is to be kind. 'The quality of mercy is not strain'd, It droppeth as the gentle rain from heaven/Upon the place beneath.'[1] You must always be merciful, and thoughtful, as you deal with those lovely students."

Then, with the soft sound of that gentle rain from heaven, she disappeared.

CHAPTER 4:
Acquiring Knowledge

Richard Swenson, in his lovely book, *More Than Meets the Eye*, states clearly the revelations in store for a student of Anatomy: "Every human body is a miracle exceeding comprehension. The complexity and dimensions involved are staggering to the mind, straining our abilities to apprehend the grandeur. Yet if we make the effort to reach for the meaning behind the biology, we will find ourselves edified on several levels.

First, we will see anew the *power* and *precision* of God. When we think of the majesty of God, we don't have to look through a telescope to the distant recesses of the universe – we need only look in the nearest mirror."[1]

Every August

One of the most exciting things in my day-to-day life is starting a new academic year in August. I've been doing this kind of work since before you were born, and it still gives me an exhilarating feeling to anticipate the beginning of my classes. It feels like when I was on the basketball team in high school, "riding the pine" as a proud member of the third string of players. When the game was out of reach for one team or the other, the coach would look mournfully down the bench and say, "Forbes! Go in for So-and-so!" It would give me butterflies in my stomach, but my entire life at that time centered around those few moments I spent in the game. It's the same today, but nowadays it relates to my class at school. The sensation I felt in my stomach as I anticipated playing basketball is the same sensation I get each summer when I anticipate the beginning of school. I don't think I'm unusual in this regard; I have colleagues who have told me that they feel the same. We simply have the best gig in the world.

Here's one exciting part, for me: I always try to memorize the names of my students *before classes start*. I use flash cards for this exercise. Most professional schools put out a composite sheet of photographs of all the incoming freshmen, and I construct flash cards and try to memorize the names. When my wife Gerri taught the course with me, we'd compete to see who could learn all

125 names first. We'd each make our own set of flash cards for this purpose. Then, when the first laboratory session came around, it was fun to surprise the students by addressing them by name. It was a little creepy for them, but they were impressed with the fact that I "knew" them at that stage. You could almost hear them wondering, "What *else* does he know about me?"

The Need to Review

I needed to review those student-photo flashcards periodically throughout the first year. During those reviews, I would write notes on the flashcards, as I learned more about each student: their home town, health professionals in their extended family, grades in my course, and the like. Later on, when those students were in their second, third and fourth years of professional school, when I was no longer giving them grades, I would occasionally encounter them in the halls or elsewhere, and a nice friendship would develop. But in making and augmenting flash cards of my students, I was practicing a major thesis of this book: learn the names (of the students, or of the anatomical structure) first, and then add information to each name, like layers on an onion, for a well-rounded understanding of the subject.

Then I review them periodically. In learning about the students, and in learning Anatomy, I find that I need to review faithfully. If I don't review the details of, say, the layers of the foot, that information will diminish more and more in my memory, like a frog's heartbeat in winter, until it's almost non-existent. Anatomy builds on itself, and, if you don't review regularly, gaps in your learning will present serious problems down the road. In Bill Hayes' excellent book *The Anatomist*, he expresses it this way: "Like a dropped figure in your checkbook registry, this error only compounded with time."[2] So make reviewing part of your learning protocols. If you don't, your efforts to catch up will take on Sisyphean qualities.

Learning dozens of names from flashcards is a daunting task, and it takes a great deal of work. Early on, I looked hard at the photograph and repeated the name over and over. That didn't work well, for me. Sad to say, I found that if I tried to memorize the names without some system, some *method* of learning them, the names would simply become a thicket in my head. So I began to use the "Baker/ baker paradox."

The Baker/baker Paradox

Over the years, I found that the best way for me to memorize the names was with the technique outlined in the second chapter of *Moonwalking with Einstein*. In that book, Josh

Foer speaks about the "Baker/baker paradox." "The paradox goes like this: a researcher shows two people the same photograph of a face, and tells one of them that the guy is a baker and the other that his name is Baker.

A couple days later, the researcher shows the same two guys the same photograph and asks for the accompanying word. The person who was told the man's profession is much more likely to remember it than the person who was given his surname."

Foer goes on to say, "The secret to success in the names-and-faces event – and to remembering people's names in the real world – is simply to turn Bakers into bakers - or Foers into fours. Or Reagans into ray guns. It's a simple trick, but highly effective."[3]

Now I ask myself: why not use that same technique when memorizing the students' names? If I can associate some characteristic with something in the name, *no matter how bizarre*, it enhances my memory tremendously. Reagan becomes ray guns, as Josh Foer says. And I suppose Forbes becomes four "B"s.

Now, if that's true (and it is) why not use the same technique to learn the myriad relationships between structures of the human body? When you study the human shoulder, you will learn that pectoralis *major* muscle and teres *major* muscle both attach to the lateral side of the humerus bone. You will learn that *latissimus* dorsi muscle also attaches to the humerus bone, but what is the relationship of the three bones' attachments? If you use a little imagery and imagination, you may conclude that a beautiful lady ("lady" = "lati-") is standing between two military men (each of whom has attained the rank of major). "Lady-issimus is located between two majors," get it? If it's slightly goofy like this, that makes it even easier to remember. This relationship has functional implications, and is important: the insertion of latissimus dorsi lies between the insertions of pectoralis major and teres major. This is one way to learn one of the endless anatomical relationships in such a way that you will remember it. These kinds of tricks can be a helpful alternative to learning thousands of facts by rote memory. The latter is akin to being crushed by a ton of feathers.

Many people consider the study of Anatomy a matter of learning great numbers of disjointed facts. I have a different approach: learning Anatomy is actually learning *relationships* between structures. There are techniques that can be helpful in memorizing those relationships. At the front of your thigh, in the inguinal region near where your lower extremity meets the abdomen, there are quite a few structures passing from one to the other, aren't there? Let's look at your anatomy atlas and consider in particular the femoral nerve, artery and vein, and the lymphatics. How are they related to one another? This is important; someday you'll be looking for a pulse in this area, and you'll be looking for femoral artery. Is the femoral *nerve* most medial, or is it the *vein* that's most medial? Or the artery? And where is the lymphatic drainage? It can make quite a difference if you're planning to be a healthcare professional.

There are techniques for learning and retaining those relationships in this particular instance. Some people remember the pseudo-word "NAVL" (pronounced "navel"), citing the order from lateral to medial: **n**erve, **a**rtery, **v**ein and **l**ymphatics. Other people simply see the inguinal ligament running uphill from medial to lateral, and say, "A van runs uphill." They are thus reminded that the inguinal ligament runs uphill, and the structures coursing under it are in this order: **v**ein, **a**rtery and **n**erve. The practice of using the pattern of first letters of words in a series is a form of "mnemonic," which is discussed later in this chapter.

Familiarity vs Knowledge or Wisdom

How many times have you looked at the backside of your hand? Thousands of times, I bet. Can you draw the venous pattern from memory? Not likely. Familiarity doesn't mean wisdom, nor does it supply knowledge. If you read a passage from a textbook over and over without reflection, or if you read it so many times that you have it memorized, that doesn't mean you'll be able to recall it in a week. Ergo:

1. It's critical to *not* listen or read passively. Look away from your resource frequently. Repeat the concept in your mind. Reflect on what you've read. Ask yourself questions. Draw a sketch of something relative to the passage. There have been studies done with people who play several games of chess simultaneously: how do they remember the position of the pieces on the board? Often, the key is to look away from the board for just a few seconds and reflect. You'll see good memorizers do this. They don't just stare at the figure in the atlas. They look for a while, and then they look away while they think back about what they've seen.
2. Write notes. Write down the questions you pondered in the paragraph above. If there's something on the atlas plate that's brand-new for you, write a note about it, and add it to your other notes. You'll study those notes periodically, later. This concept is discussed in detail in the chapter entitled, "The Forbes Method of Anatomy," beginning on page 82.

3. Try to name the two or three most important concepts. Remember, if everything is important, then nothing is. And consider which concepts are being presented in lecture or lab, so you'll know what concepts the lecturer or lab instructor thinks are most important.
4. Put the concept into your own words. After you do, compare what you said with what's written in your source, and fill in the gaps. If possible, repeat the concept aloud to another student, or, better yet, to an instructor. "This is my understanding: the chorda tympani nerve supplies special sensory innervation and secretomotor innervation, but somatic sensory innervation is supplied by lingual nerve, on which chorda tympani hitchhikes…Is that correct?"

The Derivation of Words

It's widely acknowledged that a good technique for remembering anatomical names is to educate yourself on the *derivation* of the words used in naming the structure. When you learn the name "salpingopharyngeus muscle," you will learn that the "salpingo" part of the word comes from the Greek word meaning "trumpet."[4] Early anatomists thought the orifice of the auditory tube had the appearance of the bell of a trumpet; take a look at *Netter 5th ed.* 64, entitled "Fauces," and observe the cartilaginous trumpet-like structure on the lower right. Unpacking the name of salpingopharyngeus muscle tells us that it runs from the "salpinx" to the "pharynx." The muscle called salpingopharyngeus clearly does that, in the drawing, and on your donor body. Usually, if a muscle's name is a combination of two structures, you can assume that muscle runs from one of those structures to the other.

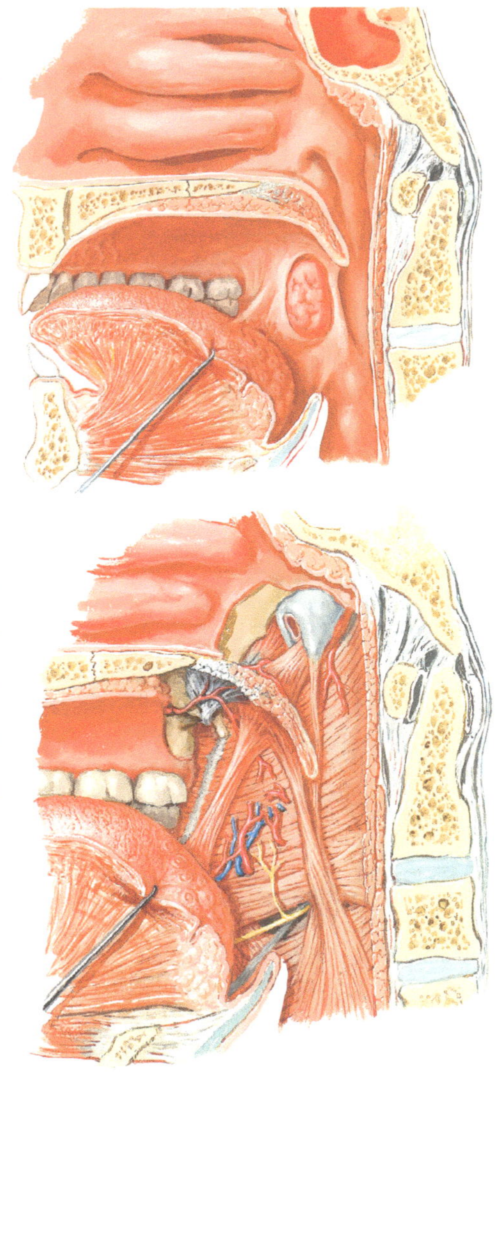

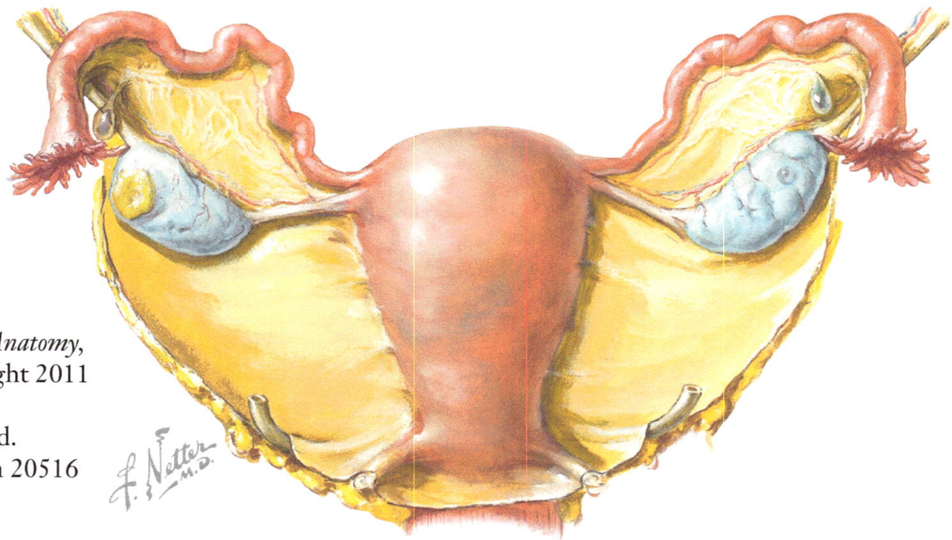

Netter Atlas of Human Anatomy, 5th Edition 355. Copyright 2011
Elsevier Inc
All rights reserved.
www.netterimages.com 20516

Armed with these facts, you will be ready when you study the pelvis: in the course of your study, you'll see the word "mesosalpinx" (*Netter 5th ed.* 355, entitled "Uterus and Adnexa"), and you will understand that this mesentery ("mes-") extends to the uterine (fallopian) tube, which resembles a trumpet ("salpinx") on its distal end. Take a look at the drawing; does the end of the uterine tube look like a trumpet to you? Early anatomists thought that it does. Knowing the derivation of the word will help you immensely in remembering its meaning.

Remembering the term "salpinx" when we encounter it in other parts of the body is a little bit like a technique Josh Foer calls "chunking." "Notice that the process of chunking takes seemingly meaningless information and reinterprets it in light of information that is already stored away somewhere in our long-term memory." He goes on to say, "…when it comes to chunking – and our memory more broadly – what we already know determines what we are able to learn."[5]

Your Studying Venue

Whether you memorize facts, learn relationships, or use the derivations of words, you will need to remember a great deal of nomenclature in your journey through the human body. Even with all these memorizing devices, sooner or later, you may simply need to find a quiet corner of the library, put a catheter in, don some noise-cancelling headphones, and spend several hours sitting with your head down, studying your textbook or atlas. My experience is that this work is best practiced in the library, deep in the stacks, alone. I'm also persuaded that if you do it at Starbucks, it may simply be a velleity.

Don't make this unpleasant. Anatomical information is valuable, and you will use it throughout your life. There will come a day when the workings of that mesosalpinx will have a profound effect on you and your family. In this day and age, perpetual learning is a necessity. A goose learns nothing, and wakes up every day to a whole new world. Don't be a silly goose. When you get tired of learning, you'll be tired of life.

The Limbic System

Jill Bolte Taylor, in her wonderful book *My Stroke of Insight*, describes the experiences surrounding her stroke that she suffered in 1996. Dr. Bolte is a neuroscientist, so her insights on brain function are particularly illuminating. The first half of her book is one of the best explanations of the basic working of the brain that I've read. It's the kind of book you could read on a weekend, and you'd be a great deal smarter on Monday. I'm happy to report that Dr. Bolte Taylor completely recovered from her stroke. Here's some information taken from her book that is germane to our work:

> The limbic system functions by applying an affect, or emotion, to information streaming through our senses. Because we share these structures with other creatures, the limbic system cells are often referred to as the "reptilian brain" or the "emotional brain." When we are newborns, these cells become wired together in response to sensory stimulation. It is interesting to note that although our limbic system functions throughout our lifetime, it does not mature. As a result, when our emotional "buttons" are pushed, we retain the inclination to react to incoming stimulation as though we were two-year-olds, even when we are adults.
>
> As our higher cortical cells mature and become integrated in complex networks with other neurons, we gain the ability to take "new pictures" of the present moment. When we compare the new information of our thinking mind with the automatic reactivity of our limbic mind, we have the ability, as adults, to reevaluate the current situation and purposely choose a more mature response.

Dr. Bolte Taylor goes on to say, "It might be of interest to note that all of today's 'brain-based learning' techniques used in elementary through high school capitalize on what neuroscientists understand about the functions of the limbic system. With these learning techniques, we try to transform our classrooms into environments that feel safe and familiar. The objective is to create an environment where the brain's fear/rage response (amygdala, part of the limbic system) is not triggered. The primary job of the amygdala is to scan all incoming stimulation in this immediate moment and determine the level of safety. One of the jobs of the cingulate gyrus, another part of the limbic system, is to focus the brain's attention.

> When incoming stimulation is perceived as familiar, the amygdala is calm and the adjacently positioned hippocampus is capable of learning and memorizing new information. However, as soon as the amygdala is triggered by unfamiliar or perhaps threatening stimulation, it raises the brain's level of anxiety and focuses the mind's attention on the immediate situation. Under these circumstances, our attention is shifted away from the hippocampus and focused toward self-preserving behavior about the present moment.
>
> Sensory information streams through our sensory systems and is immediately processed through our limbic system. By the time a message reaches our cerebral cortex for higher thinking, we have already placed a "feeling" upon how we view that stimulation – is this pain or is this pleasure? Although many of us may think of ourselves as *thinking creatures that feel*, biologically we are *feeling creatures that think.*"[6]

Dr. Bolte Taylor seems to suggest that we find a quiet, *safe* place (like the stacks in the library), where our amygdala is not triggered, for maximum production from our hippocampus.

Mnemonic Plague

Some of the lists and relationships you'll encounter in your study of Anatomy will, at first, appear to be nonsense syllables. Memorizing great numbers of names and nonsense syllables without a game plan is extremely difficult, and feels like pushing a grand piano through a peat bog. In addressing this problem, a universal technique in memorizing names and lists is the use of mnemonics. Learning lists by using mnemonics can be the difference between folding a flat bed sheet and a fitted bed sheet. Anatomists develop a phrase, or a verse, in which the first letter of each word in the phrase or verse corresponds with the first letter in the list they're trying to learn. "**O**n **o**ld **O**lympus' **t**owering **t**ops, **a** **F**inn **a**nd **G**erman **v**iewed **a** **h**op" is the classic for learning the names of the twelve cranial nerves. It worked well, and had an incantatory quality, but a commission named Terminologia Anatomica, tasked with naming anatomical structures, has recently changed the official names of some of those structures, and sadly the old phrase doesn't work any longer. Now we say, "**O**n **o**ccasion **o**ur **t**rusty **t**ruck **a**cts **f**unny. **V**ery **g**ood **v**ehicle **a**ny**h**ow."

Now there's this: is each cranial nerve **m**otor, or **s**ensory, or **b**oth? Considering the 12 cranial nerves in order: "**S**ome **s**ay **m**arry **m**oney **b**ut **m**y **b**rother **s**ays **b**ig **b**rains **m**atter **m**ore."

My *New Oxford American Dictionary, 3rd edition*, defines "mnemonic" as simply "a device, such as a pattern of letters, ideas or associations that assists in remembering something."[7] It comes from the Greek: "*mnemon*," which means "mindful." Usually, a mnemonic takes this form: the first letter of each word in the list, made into words in a phrase. The list "**T**om, **D**ick, **an** (sic) **H**arry" is a useful reminder that the important structures surrounding the ankle are ***t***ibialis posterior, flexor ***d***igitorum longus, posterior tibial ***a***rtery, tibial ***n***erve, and flexor ***h***allucis longus. You will see this pattern, and you'll appreciate the mnemonic, when you encounter *Netter 5th ed.* 517, entitled "Tendon Sheaths of the Ankle."

Regrettable but true: it sometimes seems to be human nature that the more off-color the phrase is, the easier it is to remember. When I was a student, we learned the bones of the wrist (carpal bones) by memorizing "Never lower Tillie's pants; Mother might come home." Thankfully, this mnemonic no longer works, because anatomists have changed the names of some of the carpal bones since I was in school. (Perhaps it was the salacious nature of this mnemonic that prompted the anatomists to change the names of the bones, but I don't think so.) Nowadays, you'll get along much better learning the carpal bones (*Netter 5th ed.* 444, entitled "Bones of Wrist and Hand") with a new mnemonic, only slightly less off-color: "Some lovers try positions that they can't handle."

Now, consider this: the problem students run into is that they are so umbilically tied to their mnemonics that they show up at the exam, or in the lab, having learned assiduously the names of the bones (using the mnemonic), but having neglected the application of those

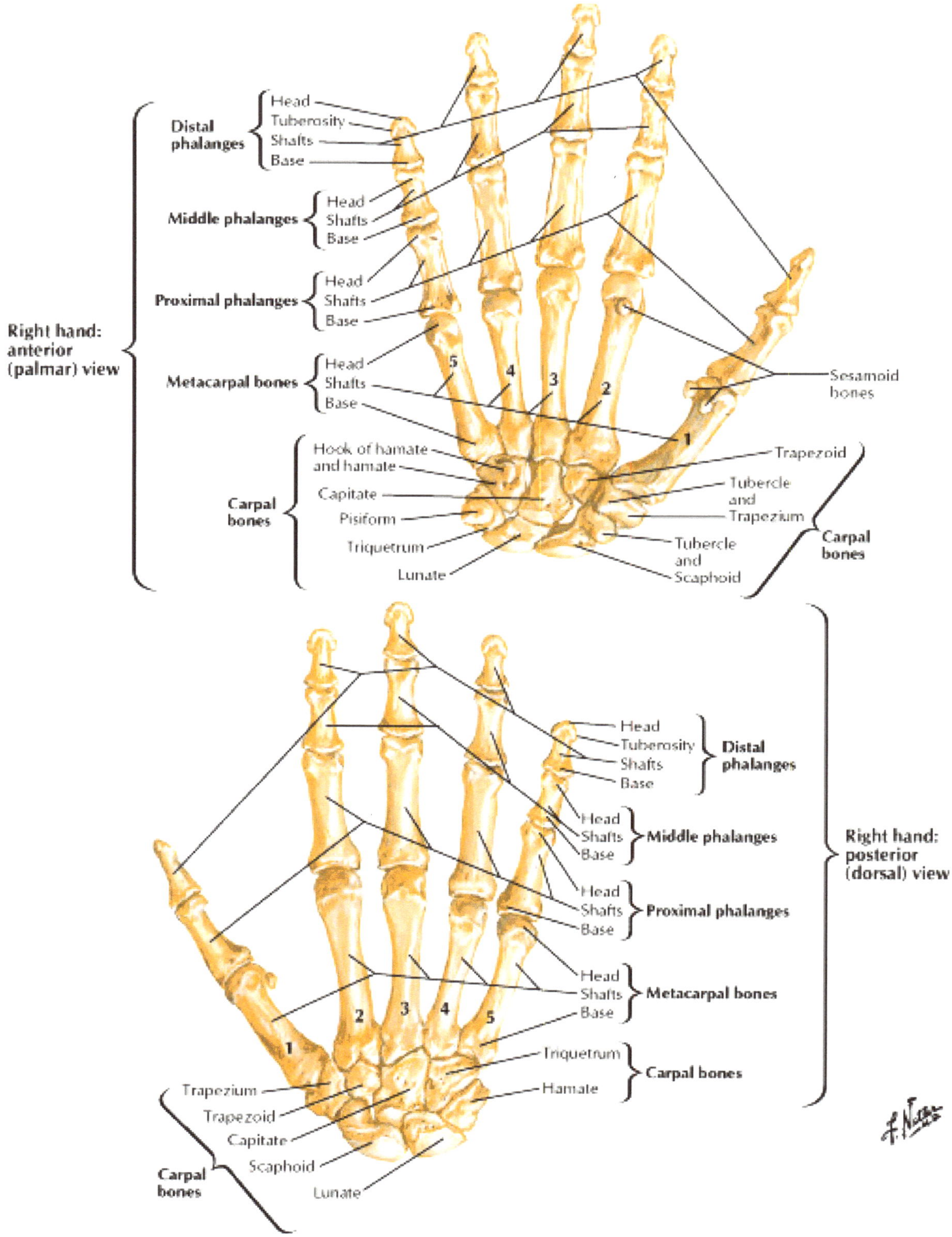

Netter Atlas of Human Anatomy, 5th Edition 444. Copyright 2011 Elsevier Inc

names to the bones at hand. Try this: look at the drawing above (*Netter 5th ed.* 444). and use the mnemonic to name the bones. In naming the bones, should you start at the medial side, or the lateral side? Should you start at the proximal end or the distal end of the wrist? If you're going to use the mnemonic above, you need to remember the following facts as you tick off the names of the carpal bones:

1. You should be looking at the *anterior* (palm) side of the hand.
2. You should be starting at the *proximal* (nearest the elbow) row of carpal bones.
3. You should be starting at the *lateral* (thumb) side.

There. Now the statement that "Some lovers try positions that they can't handle" works very well. But without the three provisions written above, the mnemonic is useless.

Here's another example: when learning the muscles that keep the back erect (the "erector spinae muscles"), many students learn the mnemonic "**I** **l**ove **s**paghetti." The first letter of each word in the mnemonic corresponds to the first letter in each of the three erector spinae muscles: **i**liocostalis muscle, **l**ongissimus muscle, and **s**pinalis muscle. Easy. Check it out on *Netter 5th ed.* 169, entitled "Muscles of the Back: Intermediate Layers." Now, ask yourself this: are we naming the muscles from lateral to medial, or medial to lateral? If you don't know, the mnemonic isn't going to help you much on an examination.

Mnemonics can be helpful, but, like donuts, they can be overdone. Some students have what I call "the mnemonic plague." By that I mean this: while a mnemonic helps a student to bring nonsense syllables back to their consciousness, it seldom helps their *understanding* of the anatomy. A different (and in my estimation, better) way to learn the three muscles in the example above, is this:

Spinalis muscle's name tells you it's closest to the spine.
Iliocostalis' name tells you that it spans from the ilium to the costals (ribs).
Longissimus muscle is the very long muscle, between the other two.

It's possible to get so involved in achieving something that you forget what it is that you're achieving. Students run the risk of clearly remembering each word in *the mnemonic*, but profoundly forgetting *what each word stands for.* I want you to have *knowledge* of anatomy, but I also want you to have *judgement* about it. William Penn, in a book of philosophy and proverbs called *Some Fruits of Solitude, wrote:*

> *Knowledge is the Treasure, but Judgement the Treasurer of a Wise Man. He that has more Knowledge than Judgement, is made for another Man's use more than his own.*[8]

The Changing Role of the Healthcare Practitioner

Remember: much of the information you will handle in your clinical practice is available to the average citizen patient. That patient can get an immense amount of data regarding his or her health on the internet and in several modes of media. It's available to the patient to read, and even to learn, but here's the thing: a lay person is prone to misinterpret it, take it out of context, or misunderstand it. What they need from you is the wisdom to interpret it, to put it in context, and to understand it.

In today's technological terms, some patients will show up in your practice with tons of information in their "hard drive." The reason they consult you is because your professional training has endowed you with a better "processor."

When I was your age, healthcare practice was a matter of treating an extant disease: diagnosis, treatment plan, treatment. Nowadays clinicians spend more and more of their time working toward minimizing the effects of risk factors in a healthy patient. The successful practice of these new paradigms requires equal parts of knowledge on the one hand, and wisdom and experience, on the other.

The clinician knows how to access good information, how to evaluate it for veracity, and how to apply it. In other words, the healthcare professional has gone from being the library of medical information to being the librarian of that medical information.

Flash Cards

The use of flash cards is a time-honored method of memorizing facts. As I wrote earlier, I use flash cards myself, when memorizing the names of my students before the advent of the course. I take the composite photographs of the class that the school publishes, and make an individual card for each student in the class. In some applications flash cards can be very helpful.

I think the greatest drawback to flash cards is that it takes so long to make them. I firmly believe that my students often make *flash cards* when they should use the time to make *notes*, as described in Chapter 6: The Forbes Method of Anatomy. In that chapter, I'll discuss some of the work of Dr. Armin Kamyab, who has written an excellent little book entitled *How to Study in Medical School*.[9] I commend this book to you: it's less than a hundred pages, but it is powerful. One of the primary theses of the book is an encouragement to the student to write his or her own notes, from the Notespaks, assignments in the textbook, and during lectures. Prepare them and augment them *every night before you go to bed*, and then use those notes to study at other times, like on the weekend.

My recommendation is not to quit the use of flash cards; they can be excellent in some applications. However, as a primary means for mastering the concepts in Gross Anatomy, I think taking notes is superior.

In the instances where you do employ flash cards, I would make them for specific information, and use them judiciously. As you know, flash cards are taking on new forms nowadays. There are apps for your digital device that can display your cards, making them readily available 24/7. You can review the cards between classes, or *during classes* (not mine, please), or while riding (not driving) in a car. You can download commercial flash cards that pertain to your anatomy class, onto your phone or computer, or you can make your own. Flash card learning is going to a whole new level with the advent of computers and the internet.

Some apps present a psychological technique called "spaced repetition," where the subject is asked to remember a certain fact with the time intervals increasing each time that fact is presented or said. Flash cards containing new information are presented to the learner more frequently than flash cards already studied and mastered. The same goes for flash cards that contain difficult, or challenging information: they come up more frequently. A flash card containing previously mastered information comes up just frequently enough to keep the information reinforced in the student's mind. Repeated exposure to the information is obviously a good way to reinforce that information in your mind. It makes sense to repeat the more difficult flash cards more frequently, and the easy ones less frequently.

Benefits received from these techniques are the results of applying technology to the concept of those old-fashioned flash cards that you and I used as far back as elementary school. Nowadays, with the use of technology, there are many uses of flash card that can make your phone a great source of information, and it's as close by as your elbow. However, I would caution against using them as your primary learning tool. That primary learning tool should be taking, transcribing, and reviewing *your own notes* (cf. page 68).

Student Life and Interest Groups

In most professional schools, there are groups of students who get together on a more-or-less regular basis, because they share an interest in a particular aspect of their study. At one school where I worked, there was a very active Pain Interest Group (PIG) that met regularly and it was common to hear someone say, "Are you going to PIG today?"

I recommend that you find a group that focuses on something that excites you, and join early. Be active in the group; head it up, if possible, but be careful not to overdo it. If you join the Geriatric Interest Group, and discover that you don't like elderly folks, switch.

In some contexts, the cadaver table team forms an excellent study group. For the duration of your time together, members of the group can contribute a great deal to each other's education.

Take a look at *CHAPTER 6: The Forbes Method*, and read *"Forbes Method of Solidifying the Information in Your Brain: the best way to learn is to teach, especially at the cadaver table and in study groups."* In that chapter, there are several insights into this invaluable approach to learning Anatomy.

Of course, there are endless groups, other than study groups, getting together on any college campus. Some of them can act as a helpful diversion from the rigors of study. Try out for the University Choral Group, join the Reptile Club, or sign up with an intramural ice hockey team. The opportunities are endless. However, the most important group you can join while in professional school is a study group.

Take Advantage of Your Family

When you head home for Thanksgiving vacation, be sure to share with your family what you've been learning. I recommend it: take blood pressures on your mother and father. Describe the median cubital vein and the process of venipuncture. Listen to heart sounds in your siblings, and tell them about the "lub" and the "dub." Study the idiosyncrasies of your family's faces, and analyze your grandfather's gait: why does he lean so to the left when he takes a step? You will find your family a treasure trove of interesting anatomical data.

And here's the thing: they will love it. They are so proud of you that they will be eager to get insights into your education in this way. They've been curious about what you've been learning, and they can't wait to tell their friends at church that you "examined" them while you were home. Teach them a little. It's fun, and it's a valuable opportunity to apply what you've been learning to some very willing patients.

One caution: resist the temptation to give medical advice or anatomical diagnoses to your family or their friends. During this exercise, if you discover something that isn't quite right, give the advice that a physician visiting their home would give: "You should have that looked at." Don't diagnose or dazzle them with your newfound expertise. Believe it or not, in most cases you're not quite ready yet to do so.

In 1745, George Washington wrote out by hand over a hundred *Rules of Civility and Decent Behavior in Company and Conversation*, presumably as part of an exercise in penmanship assigned by George's schoolmaster. The 38th of his many rules is the following: "In Visiting the Sick, do not Presently play the Physician if you be not Knowing therein."

Lots Going On

Think about this: right now, while you're reading, there are thousands of processes going on in your body, without any mental effort or conscious aforethought on your part. Your gut is moving your lunch along, taking from the contents and giving back all the while. Your heart is selecting the appropriate pace and stroke to meet your needs at present. You're blinking, you're breathing, without considering it one bit. The blood in your veins is finding its way back to the heart, largely against gravity, propelled upward by some mysterious force. Your boredom with this tedious book prompts you to yawn, requiring you, for a reason that *nobody knows*, to stretch your mouth open wide. Your eyes are focused on the appropriate distance, and accommodated to the light.

How are these legions of processes managed? The mind boggles. Your body just goes about doing a thousand processes, making a thousand adjustments, and managing a thousand relationships. Aren't you curious about it? Good. So am I. Let's keep going.

CHAPTER 4A:
Discourse with the Donor

She came again last night. She woke me up, her voice sounding like she was down a well. Actually, I was grateful that she came because I was having my recurring nightmare. In the nightmare, I arrived late to my lecture, and found myself completely unprepared. I had left my PowerPoint presentation back in my study; in any case I had not reviewed it, and had no idea what it was about. My shirt was buttoned wrong, and the fly on my pants was unzipped. Only a few students were in attendance, and the only people paying attention were my colleagues, sitting in the back row, incredulous. It was a tremendous relief to be awakened with the realization that it was all a dream.

"Hey, wake up! Rise and shine!"

I slowly opened my eyes, like a garage door going up. "I may rise, but I refuse to shine. How can you be so cheerful? It's 3:45 A.M."

"Why *wouldn't* I be cheerful? I've escaped your veil of tears, and I'm educating some very smart people every day in the Anatomy Lab. Get out of bed. Nietzsche said 'that what does not kill us makes us stronger.'" She nodded her agreement with her own comment. Her appearance reminded me that the class had recently studied the thorax: just one lung remained in her chest. On the hollow side, the lung was gone and the phrenic nerve had been preserved, draped down across her chest like a footbridge in the Himalayas.

I said, "Nietzsche also said 'One must still have chaos in oneself to be able to give birth to a dancing star. …' What on earth does *that* mean?" I thought for a moment, and then said, "You've certainly got chaos going on in yourself these days." In the gallimaufry inside her chest cavity, my eyes were searching for ligamentum arteriosum, without success. "Still… you should take satisfaction from the fact that you're having a lifelong impact on the five people who work at your cadaver table."

She pounced on that. "What? Do you know what you're saying? *Five* people, you say?"

I tried to be patient. "Yes, Monica, James, Rudy, Ellen and Elizabeth. You know, the medical students who took your lung out yesterday."

"Habibi, do you *really* think that's the extent of my influence?" She had the longsuffering look of a woman trying to teach a cat to talk. "Let's think about this. My five people will have five separate medical practices, of one sort or another."

"I think Ellen aspires to become an Ob-Gyn doc," I offered, trying to be helpful.

"O.K., an Ob-Gyn doc. How many patients do you suspect Ellen will see in her thirty-year career?"

"I don't know… I hadn't really thought about that…"

"Thirty thousand!" she shrieked. "Thirty thousand people will benefit directly from Dr. Ellen's practice, and she'll give those patients advice, and those patients will pass that information along to their bridge group, they'll tell that advice to their Bible Study group, they'll tell their children…" It was a sentence that was not coming up for breath. "…and their parents, dozens of people…" Finally, she said, "We're approaching a big number, here."

"Indeed. I tip my hat to you," I said, lifting my imaginary hat. "You're taking credit for the information she'll give to her *family?*"

"You bet. Do you think she would have learned about the miracle of conception nearly as well, had she simply read it in a friggin' *book*?"

I was beginning to find my position untenable. "No… I suppose not."

"No. And I liked the way you described fertilization in the lab today. You said the ovary tosses an ovum from its surface out into the abyss…"

I jumped in. "…where it is caught in the fimbriae like a second baseman catches a fly ball. Then the ovum travels down the Fallopian tube…"

She interrupted me. "Where she meets a nice young man coming from the opposite direction! They get married in the Fallopian tube."

I said, "But they don't want to raise a family there, so they travel down the tube to the uterus, settle down and have a baby together. Who says Anatomy isn't romantic?"

"When Ellen describes to her patient the fertilization that occurs in the Fallopian tube, whose Fallopian tube do you think she's describing? Mine! She'll never forget me. Never. And, by extension, neither will her patients. *That's* why I signed up for this duty."

"You signed up for this?"

"Yours truly. I had the greatest respect for my doctor."

"Your *doctor*? But your doctor must have failed you, in the end. You're dead," I reminded her. Then, backpedaling from what I'd just said, "Did you have an event that prompted you to give your body to science?"

"Nope. Just thought about doing some good for some other people." She sat on the side of my bed, with the sound of a wind chime in a soft breeze. "I had been suffering with depression for a long time, anticipating my impending doom. One day my pastor told me the best thing one can do to fight depression is to help someone who is not in a position to requite your kindness."

"So you're telling me you were prompted to sign up to be a donor body because you wanted to impact humanity for the better?"

"Of course. I'm not a hussy, you know." She whirled around and headed for the door. I may be mistaken about this, but I believe she shook her partially dissected booty ever so slightly, as she left the room.

CHAPTER 5:
How to Study Gross Anatomy

"The whole art of medicine is in observation." Sir William Osler

Sir William Osler has been called "the father of modern medicine," because of his radical ideas, like bringing medical students out of the lecture hall for bedside clinical training. He established the first residency program for specialty training of physicians, and was one of the four founding professors of Johns Hopkins Hospital. He took the ancient maxim above, as his teaching model. The maxim is never more applicable than in the anatomical side of medicine.

Anatomy is Special

In Anatomy we use a variety of experiences to improve our study. Bradley R. Staats, in his book *Never Stop Learning*, appreciates that approach. "Using variety to slow ourselves down makes us more likely to do what we should...Variety can switch you out of automatic mode so that you see what is happening around you and make conscious choices to learn."

Dr. Staats goes on to say, "Variety offers another advantage for learning: it is motivating. Engaging in different activities helps overcome boredom... When we engage in a variety of activities, we are stimulated rather than disengaged out of boredom. We stay motivated and are more likely to keep learning."[1] Anatomy students engage in just such variety.

The New York Times September 6, 2010, states it this way in an article entitled "Forget What You Know About Good Study Habits" by Benedict Carey: "Forcing the brain to make multiple associations with the same material may, in effect, give that information more neural scaffolding. What we think is happening here is that, when the outside context is varied, the information is enriched, and this slows down forgetting."[2]

In Anatomy classes, we've been doing that for generations. When learning Anatomy, we are blessed with those "multiple associations with the same material" in a vast array of contexts: books and journals, anatomical atlases, plastic models and plastinations, bones and skeletons, radiographs and other imaging, white-board drawings, study group discussions, professors' office hours, Acland videos, lectures and symposia, case studies, on-line study aids, our own bodies, and the ultimate study tool: the cadaver. There are several ways that you can take your professor home for dinner (I'd like that), like Panopto, with which you can watch a lecture x2, or even faster in the tedious parts, and proceed more slowly in the more recondite parts of the lecture. Furthermore, in most classes the material will be *tested* in several contexts: quizzes, graded dissections, formative exams, summative exams, verbal quizzing by the instructor in the cadaver laboratory, problem-based learning (PBL), and practical examinations. So you'll see many context variations in your study of Gross Anatomy.

Most new knowledge comes from new combinations of old knowledge. For example, when studying the anatomy of the heart, you'll bring a great deal of what you've already learned about muscles, nerves and blood vessels to bear on your knowledge of cardiology. Taking that experience and knowledge, together with the learning resources listed in the previous paragraphs, your task will be to meld all that new and old information into substantial concepts about *an organ that can successfully pump 100,000 times each day, sometimes for a hundred years.*

Surface Anatomy

Anatomy is special in another way: you have the opportunity to apply what you're learning to your own body. When you study the vermiform appendix, find McBurney point on yourself, in the mirror, and *apply* what you're learning, immediately and profoundly, to your own body. It's called "surface anatomy," or "living anatomy," and it is the topography of a living person.

My *Stedman's Medical Dictionary* defines "surface anatomy" in the following way: "the study of the configuration of the surface of the body, especially in relation to deeper parts"[3]. It's the study of the *outside* of the body, especially in relation to the *inside* of the body.

When you're studying great saphenous vein, which historically has been harvested for use in bypass cardiac surgery, trace, in your anatomy atlas, the interesting course of that vein, as it runs on the anteromedial surface of your own thigh. (In your *Netter 5th edition*, it's plate 473, entitled "Lymph Vessels and Nodes of the Lower Limb.") One interesting observation is this: as the vein descends from the anterior surface of the thigh, it courses posterior to the knee, and then courses again to the anterior surface of the leg*. By tracing structures on your own body in this way, *you* can pose as your first standardized patient**. It's a type of reinforcement that is not always available in studying, for example, biochemistry.

*Can you think of a reason that great saphenous vein should course from anterior to posterior to anterior again? Possibly the reason is this: if the vein ran its entire course on the anterior surface of the lower extremity, flexing your leg at the knee joint would stretch the vein severely. By taking "a short cut" behind the knee, this stretching is avoided.

**Standardized patient: a healthy person, usually hired by a school or hospital, to pretend they have a certain malady. The standardized patient prepares for the job by becoming knowledgeable about the condition they profess to have. In interactions with the standardized patient, the student learns to take an adequate history, and to perform a physical examination of a patient.

It has been my experience over many years of teaching, that students typically neglect the application of surface anatomy to the material they are learning. This is a mistake; surface anatomy is, of course, the approach that one will be dealing with in almost any clinical practice. Even more importantly, the application of surface anatomy to your study will make the topics literally "come alive," and will reinforce the all-important *relationships* between the various anatomical structures. When you study the common carotid artery, you will realize that finding it on a living person involves palpating the anterior surface of sternocleidomastoid (SCM) muscle. You can prove that relationship on your own body. The student investigator should develop the habit of mentally projecting the internal organs being studied onto the surface of a living person.

N.B.: Bear in mind that many of these projections will be slightly different in a person who is standing and a person who is lying down. Some such projections in the prone position are different from those in the supine position.

There are textbooks on surface anatomy, and in my estimation, Dr. Kenneth Backhouse's *Color Atlas of Surface Anatomy*,[4] is one of the best. It is presently out of print, but it's available, and it's a lovely book. You can purchase used copies on Amazon and other places, although they can be a bit pricey.

The Medical Dictionary

If you buy a good medical dictionary and keep it handy, it will answer 10,000 questions for you in the course of your professional education. Everything from the proper pronunciation of "gastrocnemius" to the real meaning of "interstitial." It will inform you regarding the insertion of brachioradialis, and the innervation of the stapedius muscle. If you keep yourself perpetually aware of this indispensable resource, it will save you a great deal of time and effort, in your exegesis of even the most intractable concepts in anatomy, and medicine.

Many people have vinyl records and vintage books that they've gone to, over and over for many years. Over time, those sources of pleasure become like old friends. My medical dictionary is like that. It's canonical for me, and it has the dog-eared, disheveled look to prove it.

As I write this, anyone can get a medical dictionary like mine, *Stedman's Medical Dictionary, 27th Edition*, on Amazon for ten bucks. I'm an Associate Professor of Anatomy, and I refer to it *on average, a half-dozen times a week.* Honestly, I do. I hope you will, too.

Keep Up

"Small bites are easier to digest." Work on Anatomy each and every day, at least a little bit. You can't cram in Anatomy; there is simply too much material to gulp down all at once. I would suggest that an important aspect of this work is to transcribe your handwritten notes at the end of each day. (*cf:* Chapter 6: "The Forbes Method of Anatomy," page 83)

Some people fall behind in Gross Anatomy, and nonetheless manage to pull it off. Most can't. In graduate school and professional school, you are sitting amongst a group of people who have all excelled. They've all achieved academically, just like you have. Everybody is giving it 100%. Do you really want to handicap yourself in *that* cohort? Be assured, most of those very people – those overachievers - find that they can't be successful by cramming; not at this level. Your best approach to competing in this arena, especially in Anatomy, is this: stay current, and don't fall behind the rest of the class.

Here is the preeminent truth about Anatomy: *Anatomy builds on itself.* It builds on itself to such an extent that if you're weak in, say, the anatomy of the skull, you'll be lost when it comes to learning the course of the cranial nerves. Or the blood vessels. Or the attachments of the muscles.

Anatomy builds on itself in another way, as well. To wit: if you learn well the pattern of nerves in the *upper* extremity, it will help you a great deal when you come to learn the pattern of nerves in the *lower* extremity. There are some distinct similarities. Rather than being a group of disjointed facts, one part *builds* on another, constantly, and it does so to an extent you may find hard to believe.

When you study the abdomen and learn the prevertebral ganglia, you'll realize that they are served by thoracic splanchnic nerves (sympathetic), and vagus nerve (parasympathetic). Then, when you study the pelvis, you'll realize that the inferior hypogastric plexus serves the same purpose. It is served by sacral splanchnic nerves (sympathetic) and pelvic splanchnic nerves (parasympathetic) in the same way. These correlations obviate the need to learn the material all over again; you've already learned it.

I know, *I know!* Professors in *every* discipline advise their students *not* to put off studying until the night before the exam. I think this is obvious, but this admonition is particularly germane to the study of Anatomy. (Am I biased? Probably.) Why is it so important? Because Anatomy is about relationships. Repetition, and reading the material more than once, and in different contexts, will serve you well in recognizing relationships. That's much better than one vigorous, all-night reading of the material. Your professor has carefully organized the course in such a way that you first learn the basics, and then you build on them. You learn the three divisions of the trigeminal nerve *first*, and then, in repeated visits to those divisions, you add the branches, then the terminal effects, then the autonomic "hitchhikers." Learning the trigeminal nerve all in one night, without reflection, compromises this important practice of learning anatomy gradatim.

I have taught Anatomy in a medical school, a dental school, several residencies, and undergraduate schools. I've taught in a physician's assistant school, a mortuary science school, a pharmacy school, a physical therapy school, and a nursing school. Based on this considerable experience, I can tell you: the study habits that serve a student well in undergraduate courses will bitterly disappoint you when you employ those habits in an Anatomy class in graduate school or professional school. We say, "What got you here won't get you there." Struggling students show up in my study and they boldly tell me, "Dr. Forbes, I am *not* a 'C' student." But they find themselves in the "C" category (or worse) because they have carried their undergraduate study habits into graduate or professional school.

Dude: all of us achieved in undergraduate college, and many of us did so without a great deal of effort. In professional school, it's different: everybody's smart. You will not be in Anatomy class very long before you realize you're drinking from a fire hose. The information you need to learn just keeps coming. And examinations keep coming. Waiting for that next examination is *like following a lighted fuse in search of an explosion: one never has to wait very long.* You need to select carefully what you need to know, *and then keep up with the class* in learning it.

Diversions

Of course, it's important to get away from the study from time to time. It refreshes you, and improves your overall outlook. This is not a new concept; Plato's *Phaerdrus*, written in 370 BC, states: "The mind ought sometimes to be diverted, that it may return to better thinking."

However, what happens to a great number of students in professional school is this: they see some of their successful classmates skipping class and some of their successful classmates staying up late playing poker, and some of their successful classmates going home every weekend. What the weak student may fail to understand is that the successful classmates are practicing *some* of those diversions, but no one is practicing *all* of those diversions. The weak student may choose to practice *many or all of* the diversions, and could fail as a result.

If you have to give more time to your studies, you have to give more time. You must decide if your goal is worth giving up, for a while, some activities that you desire. And my final bit of advice is this: if a particular strategy isn't working for you, *don't do it*. Ralph Waldo Emerson said, "A foolish consistency is the hobgoblin of little minds." If you do poorly on the first exam, don't wait until after you've done poorly on the second exam, to make a change. Mark Twain said, "There is nothing to be learned by the second kick of a mule." Be willing to try something else. Or try the same strategy, only try harder.

Pick one, or a few, diversions and study hard the rest of the time.

Never Studied Anatomy Before?

This may surprise you, but most students who are admitted into health professions programs have never taken Anatomy before entering their program. That goes for medical school, dental school, physician assistants' schools, and on and on. If you arrive at your program without prior preparation in Anatomy, one important bit of advice is this: learn the basics of Anatomy first. Don't bog down in details at the beginning. Get yourself a primer on the subject, a very elementary textbook on Anatomy, and learn the basics first. My favorite for studying the head and neck is *Illustrated Anatomy of the Head and Neck*, by Fehrenbach and Herring[5]. If you're studying the whole body, try using the dissection manual in your course as the means of learning the basics.

When you learn a new area of the body, pick a plate in an atlas, or the textbook of your course, and learn it well. Then, after you have that basic information solidly in your head, hang all new information on that plate. Have the basics firmly in your mind, *then* hang new information like decorations on a Christmas tree.

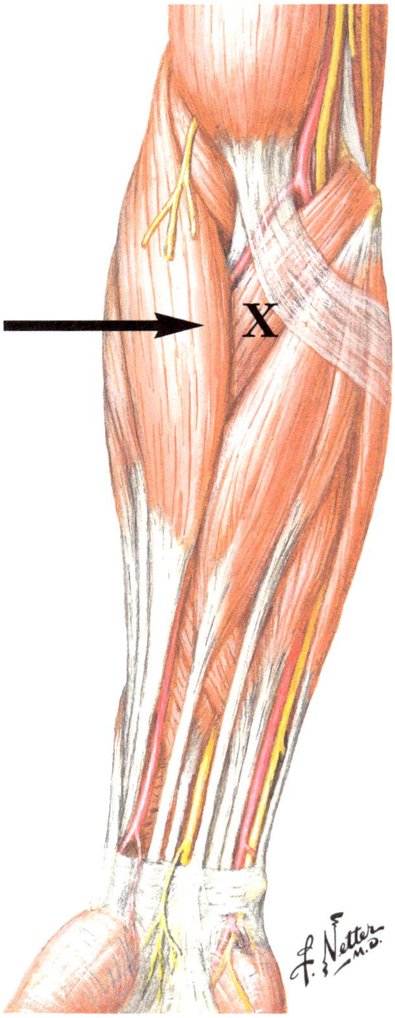

Netter Atlas of Human Anatomy, 5th Edition 434. Copyright 2011 Elsevier Inc All rights reserved. www.netterimages.com 63

Knowing Right from Left

Drawings in an anatomy atlas almost always portray the *right* side of the subject. That's the way you'll burn them into your visual cortex. When you come to a situation on the cadaver's left side, you'll need to make an adjustment. You'll have to think in a mirror image of what you introduced into your brain. That can be done, but the important thing is to remember that most atlases consistently portray the right side of the body.

For example: when you envision the brachioradialis muscle (whose name tells you that it runs from the brachium [arm] to the radius bone), you'll see it, marked by the arrow, on the *right* side of the body, as it is on *Netter 5th ed.* 434, entitled "Muscles of the Forearm (Superficial Layer): Anterior View." You'll remember that it sweeps around the lateral side of the elbow, and heads for the wrist. You will probably have your atlas's picture of that muscle firmly in your mind. However, your professor may test it on the *left* side on a practical examination.

For most people it's best to learn that right side drawing, and reverse the structures in your mind when confronted with the left side. When you're standing at a donor body on the practical examination, and you're working to recall an image, that image in your head will almost always be the *right* side, and you'll need to do the conversion in your brain. This is important and a little counterintuitive: the more firmly you have the *right side* ensconced in your brain, the more smoothly you'll be able to convert to the *left side.*

Let's take another example: this can be important if you're trying to remember the configuration of pronator teres muscle, marked with an "x," on the previous drawing, (*Netter 5th ed.* 434). This is a muscle that pronates the forearm, which is to say it takes your hand from the "holding soup" position (supination) to the "dumping it on the floor" position (pronation). The muscle's shape on cross-section is round, and the word for "round" is "teres." It's easy to remember that pronator teres muscle slants downward from an epicondyle. But is it the *medial epicondyle*, or the *lateral epicondyle?* If you have a visual perception of pronator teres in your mind, you can assume that the drawings are almost always of *the right side of the body*, and you'll realize it's the medial epicondyle.

Figure Out What's Important to Your Professor

You can't learn everything about Anatomy. I've been doing Anatomy for forty years, and I'm still learning *a great deal.* I know you have ambitious goals, and for a variety of reasons, you *really* want to do well in this course. You passionately desire to know it all. But you soon realize that, realistically, you can't learn everything. Here's a tip: if you want to know what is going to be on the test, *figure out what aspects of the Anatomy are important to your instructor.* Anatomy professors are going to test you on material that they consider important, or that they themselves are interested in. And that's naturally the material you'll find in their lectures, their PowerPoint presentations and their Notespaks. And their examinations.

How are you going to determine what's important to your professor? Start by reading the Syllabus for the course. Then scan the Objectives for the course, and the Objectives for each lecture and lab session. Then *attend* the professor's lectures and take notes. If he teaches in the Anatomy Lab, take a minute and chat with him or her. *There it is.*

Get a Book that You Don't Mind Marking Up

No matter whether you're reading a Coursepak or Notes Packet, a textbook or an atlas, be sure you have a copy that you don't mind marking up. I've seen students who photocopy pages in their textbook, so they can maintain the book in mint condition for a better price when they sell it. In my estimation, this is a waste of time. Go ahead: buy a used copy if you need to, but *mark it up a lot.* When you come to review, you'll find that your marginal notes, underlines, and highlights will save you a great deal of time.

Cross-Sectional Anatomy

Cross-sections have an important role in our study of the human body, and their use has

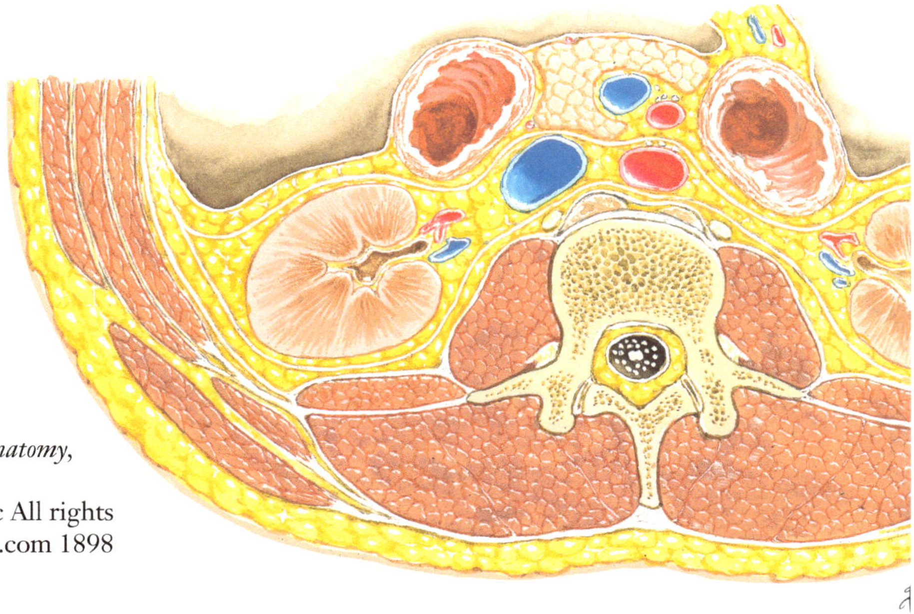

Netter Atlas of Human Anatomy, 5th Edition 173.

increased in recent years, in large part because many of the imaging techniques in today's clinical practice appear as cross-sections. Modern diagnostic techniques are often about interpreting cross sections on images.

In most cases, the professor will identify cross section images or drawings that you are required to know for the course. In addition, the recent editions of most atlases have cross sections for study. There are entire books of cross-sectional anatomy as well, like *Sectional Anatomy of the Head and Neck, a Detailed Atlas.*[6]

Many schools have acquired libraries of plastinated specimens, including cross sections of entire cadavers, for study. Plastinated specimens are actual sections of actual human bodies that have been transformed into polymers.

I advise you to leave the study of cross sections until after you have studied the Anatomy in the traditional ways. Once you have the arrangement of the structures firmly in your mind, the cross-sectional images will confirm and reinforce the arrangements and relationships of the structures that you've learned.

N.B.: Be aware that with most cross-sectional *imaging*, we are looking at the patient *from below*. Our perspective in looking at the cross-section image is an *inferior one*, as if we were looking up at the cross-section from the floor. This can add an element of complexity, since the inferior viewpoint can be counter-intuitive when it comes to telling the patient's right from the patient's left on the image.

Distractions

Are you distracted by something you need to do in your personal life that you've overlooked? If you're studying and something unrelated to the subject comes to mind, *write the distraction down* on a piece of paper, put it in your pocket, and then go back to your lucubration. For example, if you're studying the anatomy of the urinary bladder, and the study brings to mind that you need to get the Super to fix the toilet in your apartment, scratch it on a piece of paper, stuff it in your pocket or somewhere else on your person, and forget it. You'll find the note later when you clean out your pockets.

There. Now you can concentrate on your Anatomy.

Anatomical Marriages

Remember: the study of Anatomy is about each structure's relationships with other structures. Relationship, relationship, relationship. So is marriage. In some cases, anatomical structures run their course together, or parallel to each other. Most of the time, they have the same name, and that makes them easier to remember. In other cases, their two names are different. Here are some of my favorite examples:

> Lateral thoracic artery is married to long thoracic nerve. (Incidentally, they're on a date; they're going to a movie called *The Serratus Anterior Muscle*.)
>
> Thoraco *Dorsal* nerve is married to Thoraco *Dorsal* artery. They have the same last name, and their tributaries do too. (For their dates, they're fond of eating at a place called *The Latissimus*.)
>
> Spheno *Palatine* artery is married to Naso *Palatine* nerve. (She liked his last name, *palatine*, so she took it as her own.)
>
> Superficial *Temporal* artery hangs with Auriculo *Temporal* nerve. (Her old-fashioned parents are thrilled that she's taking his last name.)
>
> Cephalic vein is a confirmed bachelor; a loner. He doesn't run with an artery, nor with anybody else. His given name makes him feel like an insecure orphan; it comes from

the Latin, and means "relating to the cranium or head," and yet he's never been to the head in his life. His name is inexplicable. Possibly it's because he constantly *flows in the direction of the head.*

The Atlas

Throughout your study, you should refer without ceasing to the atlas. This book you're reading refers specifically to the 5th edition of the *Netter Atlas of Human Anatomy.* There is a plethora of atlases out there that are more than excellent. When you select the one that you like, buy two: one for the lab, and one for your home. Buy one new, and one used. Second copies are important if you have a roommate or a spouse; you will find they have very little sense of humor about a lab copy of an anatomy atlas, reeking of formaldehyde, on the kitchen table.

The Importance of the Cadaver Lab

Use the cadaver lab as your 4th time through. Use the dissection to your advantage. You've read the Objectives, attended the lecture, looked at the pictures in the Laboratory Dissection Manual, and you've perused the Notespak three times fast. Your arrival at the cadaver constitutes your fourth time through the material. In our study of Anatomy, we have the learning tool that far exceeds any other: dissection. Properly prepared, the dissector frequently finds those "Aha!" moments. Because you've prepared, you'll have a good idea of where ligamentum arteriosum is, and when you search out and dissect this amazing structure that shunted the blood away from your lungs when you were *in utero*, you will *never* (!) forget it. Many of us will retain it in our visual cortex until we die.

You will learn it better, having explored and dissected it, than you would learn it in a textbook or video. And you'll retain it better, too.

School administrators, as a rule, are not practitioners of scrupulous anatomical rigor. Believe it or not, I have heard administrators say things like, "Our students study only plastic models. They don't waste time dissecting; they use the valuable time when they would be dissecting, in the library, actually *learning* the material." For the study of Anatomy, nothing could be more absurd. Dissecting cadavers is the embodiment of all the trendy catch-phrases of today's medical education establishment: problem-based learning, active learning, students teaching students, student research, three-dimensional learning, eschewing "the sage on the stage," and on and on. It's what we do in the cadaver lab every day, and we've been doing it since Vesalius.

The Forbes Method

In the next chapter, we'll apply the principles of learning and memorizing, to Gross Anatomy in particular. There are specific, concrete approaches to learning Gross Human Anatomy that can help you in your quest. You will probably not utilize all the ideas, but utilizing some of them will help you avoid the frustration and drama encountered by some students of Human Gross Anatomy.

CHAPTER 5A:
Discourse with the Donor

Last night I read in bed until 2:00 A.M., because I just couldn't put down my new edition of *Lachman's Case Studies in Anatomy*. Shortly after falling asleep, I felt the presence of a familiar specter in my bedroom. Her voice was deep, like a double-bass-string twang.

"Hey, Doc, that student who works by himself in the evenings and on weekends must be planning to be a dentist."

I put aside my annoyance because I was interested in what she had to say on that subject. "Why would you say that?"

"Because he gets such pleasure in working by himself in the lab. He's a loner; that is a good predictor of dentists versus physicians."

I sat up in bed. I had the feeling that what she had just said could be a window into learning something new about my beloved students. "Could you please explain what you mean?"

She shook her head and tutted. "Doctor, everyone knows this: dentists are like seagulls; physicians are like quail. You've seen seagulls on the peak of a house roof: they space themselves absolutely equidistant from each other, and if a seagull flies away, everyone adjusts in order to provide maximum space between each two; if another seagull joins the party, everyone adjusts accordingly. Equal distances. Like notes on a bar of music."

"And the physicians?"

"Physicians, on the other hand, are like quail. If you're hunting in the woods, and you scare up a bunch of quail, they fly into the underbrush in every direction, but before you're out of earshot, they've grouped back up again, and that's how they like it."

"Hmmm… which is better?"

"It depends on your personality, and your constitution. Working by, and for, oneself is satisfying, but having the availability and wisdom of a group can be reassuring in the decisions one makes in any kind of healthcare."

"Evidently you've thought a great deal about this." It was a question.

"I have. On the one hand, dentists have the satisfaction of fixing something. Patient presents with a problem, and nine times out of ten, the dentist can make the patient whole again, sometimes in a single appointment." I was beginning to see her point.

"On the other hand, I've always considered the practice of medicine to be like a giant chemistry experiment: the doctor tries one medicine, and if it doesn't bring the desired result, then he or she increases the dosage; if that doesn't work, he or she tries another medicine. Get another blood test, and we'll see what your numbers are, and adjust accordingly. For the most part, it doesn't have the immediate gratification that the practice of dentistry does."

I wanted to join the conversation. "I suppose orthodontics and periodontics are exceptions to the rules. As is the practice of dermatology, in large part."

"Of course. Now you're catching on." She spoke slowly, and looked at me as a grammar school tutor looks at a slow learning child.

I said, "It's probably not an accident that those specialties are generally more lucrative than the others."

"It's such a mistake to focus on the money. You can't take it with you; I'm proof of that." She paused for a moment, apparently thinking about her particular situation. "You remember how much money Nelson Rockefeller left when he died, don't you?"

"No. How much?"

"All of it," she said. "I'm leaving now. If you want to talk more, you know where to find me." Poof! Gone.

CHAPTER 6:
The Forbes Method of Anatomy

Anyone who has struggled with Anatomy (and who hasn't?) has contemplated strategies to ameliorate this struggle. This chapter is a distillation of the learning approach that has been described in the first five chapters of this book, with particular application to our study of Gross Anatomy.

Allow me to list the twelve tenets of my own personal approach to learning Anatomy, and after I've listed them, I'll discuss each of them thoroughly (you might say, "exhaustively"). You won't adopt all of my tenets, but you might think about which ones are suited to your personal learning style.

1. Forbes Method of Preparation: skim the relevant pages in the textbook, handout or laboratory manual *before* going to class. Look at (don't memorize) one drawing in the atlas pertaining to the day's subject.
2. Forbes Method of Studying the Professor's Notespak: if your professor distributes notes for his course (as I do), read them after the lecture *three times fast.*
3. Forbes Method of Learning: during the lectures, *take notes by hand* on a sheet of paper or notebook, or on an atlas plate, and do the same as you read your textbook assignments and notespaks. Carry your notes with you, and tap them at every opportunity.
4. Forbes Method of Memorizing Names of Structures: *look away* from your work periodically; take a few seconds to reflect on it.
5. Forbes Method of Organizing the Material: learn the basics first, then add details later on. Learn well one atlas plate for the area you're studying, and then, in your mind and/or on the actual page, hang subsequent information on that plate.
6. Forbes Method of Solidifying the Information in Your Brain: the best way to learn is to teach, especially at the cadaver table and in study groups.

7. Forbes Method of Maintaining Your Attention: variety is the spice of Anatomy.
8. Forbes Method for Learning Nomenclature: break the word down, and look it up. Write what you've learned next to the term in your notes or in your anatomy atlas.
9. Forbes Method of Taking a Test: take it after a good night's sleep.
10. Forbes Method of Practical Anatomy: In the cadaver lab, if you are confronted with a structure you can't identify, find something that you *can identify*, and work from there.
11. Forbes "Tricks" in Learning Anatomy: mnemonics and the like
12. Forbes Method for Groups of Facts: rules and exceptions

Let's talk about each of these individually, and then we'll go on to some general advice about learning Anatomy.

1. Forbes Method of Preparation: skim the relevant pages in the textbook, handout or laboratory manual *before* going to class. Look at (don't memorize) one drawing in the atlas pertaining to the day's subject.

How many times have you felt confused by what the instructor is saying? A brief look at the textbook or laboratory manual before the class might have clarified where he or she is going. Have you ever wondered why one of your classmates could ask a question and the instructor responds with, "That's a great question!" Chances are that student has looked over the material in advance.

You don't have to *read* the assigned material before attending class. Just look at the pictures. Skim the reading. Read the headings, and maybe the concluding paragraphs of the different sections. This will do at least four things for you:

a. You will find areas in your reading that pique your interest. You'll *want* to read those areas because they're interesting to you. Not *all* of the reading assignment will float your boat, but *some of it* will. Even if you find no such paragraphs, you will have some idea where the professor is going, before you attend his or her lecture.

b. You will develop questions in your mind pertaining to the material. In addition to your own questions, many textbooks include review questions at the end of each chapter. These questions will make you an active listener (the best kind of listener) when you get to class, whether it's a lecture, a seminar, or a laboratory session. *When you ask a question, you carve a place in your brain for the answer*. When you carve that place in advance, you will retain that answer even better.

c. You will avoid what I call the "Ready! Fire! Aim!" approach. When students attend a lecture or a laboratory session with no prior orientation to the material to be presented, they handicap themselves in learning the concepts: they "fire!" before they "aim!." Prepare in advance of the lecture. Have some idea of what you're getting into, in today's lecture, or today's lab.

d. You will have the basics in your mind, so that when you encounter terms like "meninges" in the lecture, you will already know what they are.

I believe in this concept strongly. So strongly, in fact, that in my Anatomy course, I assign one or two atlas plates pertaining to what I'm going to lecture on. The students learn those assigned plates *before coming to each class*. At exactly the beginning of the lecture hour, I inquire if there are any questions about the material before we take the quiz. I answer all the questions, (even if in the process I give away some of the answers on the upcoming quiz). Then the quizzes are distributed, and a period of five minutes is given to complete it.

The quiz consists of one or more of exactly the same atlas plates that were assigned, but the quiz is without labels. Ten structures have arrows pointing to them, and the students must fill in the blanks, labeling those ten structures. Quizzes are collected, and then I start my lecture. Collectively, all the pre-class quizzes in the course count for 10% of your grade.

The point is this: you'll learn so much more in the lecture *if you have the basics before coming to class*. I don't have to spend our valuable time on the fundamentals, because you've already learned them for the quiz. I can therefore talk about clinical applications, amusing anecdotes, marvels of the human body, mnemonics for remembering, as well as answering questions. *The interesting stuff*. And the practice of beginning each lecture with a brief quiz has the added benefit of getting students to class on time.

Before class, look at the Learning Objectives. I can't understand why people don't do this. What are Learning Objectives? They're a list of the concepts that the teacher thinks are the most important things you should know. The administration of the school requires your professor to list them on the first page of a notespak, or the first slide of a PowerPoint presentation. Most schools will confine all exam questions to topics covered in the Objectives. If it's not in the Objectives, it's not on the exam.

Think about this: your professor can't teach all the Anatomy of the human body. There simply isn't enough time to state all there is to know about a particular subject. So he or she is constantly triaging that material to make it appropriate for your particular course. Then the professor is required to list the results of that triage as the learning objectives for the course, and present them to you. There it is.

He or she is going to tell a lot of stories, and give examples, and tell jokes, *but the meat of the lecture is required by the Administration of your school to be listed in the Learning Objectives.* So please: take a look at what you're going to hear, before you hear it. After the lecture, you might want to glance at them again, to confirm that you heard it.

As you consider the Learning Objectives, look up any unfamiliar terms. If you don't know a word, write it down and look it up later. It's as simple as that. Own a good English dictionary, and a good medical dictionary.

2. Forbes Method of Studying the Professor's Notespak: if your professor distributes notes for his course (as I do), read them after the lecture *three times fast.*

Nobody would question the idea that the professor's lectures will focus on the things that the *professor* (*the writer of the examinations, remember*) considers important. Some professors will distribute notespaks, eliminating the need to constantly scribe during the lecture. I use notespaks because in teaching Anatomy I want my students to be *looking up* at the PowerPoint slides and videos on the screen, and not *looking down*, frantically scribbling notes on paper or their laptop computers. If your professor distributes notespaks, or posts PowerPoint slides on-line, my advice is this: when you study, go through them three times fast. Ideally, this should be done after the lecture on the same day the lecture was presented.

If your particular course has notespaks in class, or on a webpage, it's logical to assume that that material is what your instructor thinks is most important. *Ergo, read that material three times fast.* For this particular material, three times quickly is much better than one time slowly. Stop and think between readings. "As you take time to think, you recognize things that you already know but haven't taken the time to understand. You also make connections between new ideas. Moreover, as you reflect and identify what you don't know, you can figure out strategies to fill those gaps… As you reflect on what you know, you recognize what knowledge you already have – and often you may discover that you know more than you thought, especially when you see the strengths and positives in your past experiences."[1]

The first time through, if there are labeled drawings on the PowerPoint presentation, or in the notespak, I study them until I can cover up the labels on all of the drawings, and say the names entirely from memory. I also skim the narrative, while underlining and making marginal notes. The second time through, I read the material through more carefully, emphasizing my notifications from the first time, underlining as appropriate. The third time through I read just the underlines, and the marginal notes. I'm not saying this is the best way *for you*, but it is definitely the best way *for me*, and I've been doing this since before you were born.

The third time you read through, you will begin to notice relationships. You'll be smiling at the structures, as you link them to other structures, and you'll understand their importance.

For example, during the third pass, you might say, "Earlier in the notes, when I learned that the bump on the humerus bone was called the deltoid tuberosity, I wondered why we would have to learn such an obscure and tiny landmark. I was almost tempted to ignore it. Now, later on in the reading, I see: it's one of the attachments of a very important muscle, the *deltoid* muscle! That must be why it's called the "deltoid tuberosity." That is a simplistic example of why we go through three times fast, instead of one time slowly. I don't know about you, but I could have missed the connection between the bony bump and the muscle, had I read the passage only one time.

These sorts of "Aha!" connections will greatly augment your learning, and they will only come with repetitive readings. Many students simply won't do this, but I promise you: on your second and third times through, you will begin to make connections (relationships! the heart of Anatomy!) between structures that you hadn't made before.

That's it. For my students who have followed this advice, it has served them well. I'm asking you to *try* it, at least once, before throwing this book in the fireplace. I hope you will resolve to read your Anatomy handouts three times fast, just as you've resolved to floss your teeth every day. Except in the case of the reading, I hope you're *actually going to do it.*

3. Forbes Method of Learning: during the lectures, *take notes by hand* on a sheet of paper or notebook, or on an atlas plate, and do the same as you read your textbook assignments and notespaks. Carry your notes with you, and tap them at every opportunity.

Dr. Armin Kamyab has written an excellent book entitled, *How to Study in Medical School*,[2] I commend this book to you. It's less than a hundred pages, but it is powerful. One of the primary theses of the book is an encouragement to the student to take his or her own notes, from the Notespaks, assignments in the textbook, and during lectures. Prepare them and augment them *every night before you go to bed*, and then use those notes to study at other times, like on the weekend. Kamyab's counsel is specific advice for medical students, but what an excellent concept for any student of Anatomy! Don't take my word for it: you should buy Dr. Kamyab's book and read it carefully in order to put his proposals to work effectively.

I recommend modifying slightly Dr. Kamyab's ideas when you apply them to Anatomy. Instead of a page of notes for a lecture or a reading, in Anatomy class I'd replace the page of notes with one or two plates from an atlas. I'd add additional information *to the plate itself*, as I discover details in my lecture notes or additional reading.

I think this is funny: when I was your age, without benefit of *How to Study in Medical School*, I practiced the notes-taking principles described in the book. That's because in those days there were no PowerPoints, no laptop computers, and certainly no Panopto. Instead, we furiously took notes, heads down, throughout the lecture. Then I'd go back home and copy my notes over, eliminating what I thought unimportant, and revising them into a concise form. It was important that I copy my notes over *that same day*, because my handwriting is terrible. I've been told it looks like someone dropped a cockroach into some ink, put him on a sheet of paper, and let him walk around awhile. Obviously, it was important to go back over my notes the same day, while the concepts I wrote about were still fresh in my mind.

One point that Dr. Kamyab makes is particularly germane to my experience. Nowadays, as a thoroughgoing professor, I spend several hours each day simply reading. And occasionally I find my mind wandering: I keep right on reading the words on the page but my mind is planning my evening, or worrying about my family. Or maybe I'm pondering whether Edward de Vere actually wrote Shakespeare's plays. Suddenly, I realize that, although I've been going through the mechanics of reading, I couldn't tell you the most basic thing about what I just read. Occasionally I do the same thing while listening to a lecture (some lectures more than others!)

What makes that germane to the study of human anatomy is this: making notes is part of what Dr. Kamyab calls "actively studying," and in my own personal experience I've found it to be essential. Have you had that experience of mind-wandering while you continue reading? (Are you doing that right now?) Notes-taking can help. I've been told that taking handwritten notes is the equivalent of hearing the material seven times.

After looking through the professor's notespak three times fast, take notes that same day, on the areas of the parts of the notespaks that you still haven't committed to memory. Take notes, also, when you read any assignments in textbooks or journals. This doesn't mean re-writing, or paraphrasing, the textbook. Keep those notes pithy. Try putting them on an atlas plate that pertains to the subject. Then carry the plate, with your notes written on it, on your person, and review it in the line at Starbuck's.

If you limit your notes to those parts of the lecture that are not entirely clear, you'll be emphasizing that particular information. By limiting your notes to areas you *don't* already know, you'll minimize the amount of time spent re-reading what you *do* already know.

Don't be a stenographer. Don't feel like you have to write down every word. That's what Panopto is for. Write down things you didn't know before, and anything having to do with those Learning Objectives you checked out before class.

Use the margins of your notes-taking to make lots of reminders to yourself: "Check this out in the book," or "What does this mean?" or "This point (or term) is confusing!" or "Dr. Forbes is a dork!" Write things like that. Reflections. Circle the words you need to look up. Give yourself plenty of space for these remarks, and you'll be surprised how much you'll put in there.

Be sure to put the date and the topic at the top of the sheet, and the instructor's or instructors' names.

Then, when you prepare for an examination, Dr. Kamyab says,

"Twenty lecture hours, eight hundred slides (at about forty slides per lecture), and ten textbook chapters have been replaced by twenty-five to thirty-five pages of the most important information you need to know and on which you will likely be tested. Better yet, you have already retained the vast majority of the material that is in your notes, simply by having written them!"[3]

As an anatomist, I make copious notes on plates from my *Netter Atlas*. I pick a plate to memorize (I call that selected plate the "Reference Plate"), and I write all over it. Learning the plate is the foundation of the building. Adding subsequent notes and understanding is putting in doors and windows. If you want to keep your atlas in pristine condition, or if you want to carry your marked-up plates with you, make a color copy of the plate to write on. Nonetheless, my preference is to write on the atlas itself. I photocopy those pages to carry with me. My atlases, therefore, would not be much good to sell on Amazon; they're so full of hen-scratching and dog-eared pages that they look like they've been dropped into a bucket of water. Important structures wear the yellow badges of my highlighter. The scribblings and notes in the margins would challenge an Egyptologist. A used copy of almost any atlas is cheap, so I recommend you buy one and go crazy marking it up.

This is important: Make connections during the lecture. Try to make connections with other disciplines you've studied. Connect concepts with others that you've learned in previous sessions in Anatomy. With your experiences outside of school. With family lore: Aunt Lynne told you she had trouble breathing during delivery of her baby, and the doctor told her it's because the anesthetic in her pelvis found its way just a little too high, all the way to the diaphragm, and the doctor said that was what anesthetized her diaphragm. Make connections with your Aunt Lynne's experience when the lecturer explains that the innervation of the diaphragm *comes from the neck*. Aunt Lynne's doctor was in error.

4. Forbes Method of Memorizing Names of Structures: *look away* from your work periodically; take a few seconds to reflect on it.

If you've ever worked on a difficult jigsaw puzzle, you are familiar with the experience of having searched for a particular piece of the puzzle for a long period of time. Eventually, a reasonable person concludes that the elusive piece is on the floor, or missing from the box, or "borrowed" by an uncle as a practical joke. It simply isn't there.

It's a good idea to give it up for a while. Have a cup of coffee, or a nap. When you return to the puzzle, chances are good that you will find the piece you are missing. It was there all the time, but you were approaching it from a different viewpoint, or under different assumptions.

When you read a concept or idea that you did not know before, look away from the book or computer, and review that concept or idea in your head. *Think about it* for a moment. Teach the concept to yourself. Do this before you continue: most people cannot reflect on what they just read while they've gone on to read something else. If you're in a hurry, you'll probably skip this step, even though it might enable you to memorize *faster.*

5. Forbes Method of Organizing the Material: learn the basics first, then add details later on. Learn well one atlas plate for the area you're studying, and then, in your mind and/or on the actual page, hang subsequent information on that plate.

I enjoy reading history, and when I do, I learn a few dates in a particular era, use them as my frame of reference, and learn all the subsequent events and culture by relating them to those dates. Pearl Harbor was 1941; the Civil War was about 1861-1865. The same idea pertains to learning one plate in each area of Anatomy. For example, when studying the elbow, I'd learn comprehensively an atlas drawing that shows *just the bones* of the elbow. Then I would set about adding information to that slide, information taken from lectures, from lab, and from my reading, gradually adding "meat to the bones."

You're not going to be able to learn all the plates in the atlas, even if your IQ is 'way over on the right-hand end of the bell curve. So what can you do? This is important: If your instructor assigns atlas plates that you must learn, the answer is obvious: learn those. If the instructor does not assign specific plates, I recommend you yourself select *one* not-too-busy atlas plate that has a large percentage of the material that is in the reading or lecture. Cover up the labels, and say the names. Concentrate on the relationships between the structures in the drawing. Learn that plate backwards and forwards. That plate will be your "gateway drug," introducing you to the topic. Then, when you read the textbook or attend a lecture, instead of making notes on a blank piece of paper, write your notes directly on this plate.

That's my bottom-line best advice for learning Anatomy: find a drawing that displays the *basic information* about the area you're studying. If it were the structures of the face, I would select the top drawing on *Netter* 5th *ed.* 24, entitled "Facial Nerve Branches and Parotid

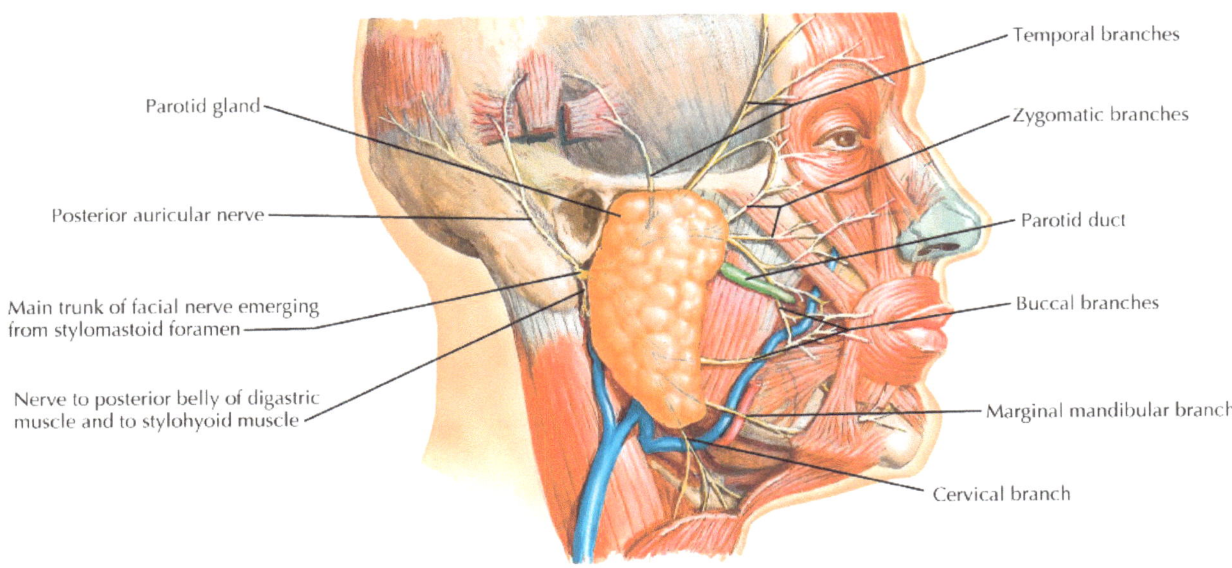

Netter Atlas of Human Anatomy, 5th Edition 24.

Gland." I'd call that drawing my "reference plate" for that region. I'd learn the labels until I could cover up the labels with a 3x5 card or my hand, and say them off reliably. I might sketch the drawing from memory, in a *very* stick-figure-like way.

O.K., now I've got the superficial structures of the face down cold. Now, when I delve more deeply into the face, I'll see that there is more detail for the same area I previously memorized. I will find it remarkably simple to hang the new information (the "doors and windows") on the drawing ("the building") that I already have in my mind.

This method gives you a *visual*, on which you can later hang the lecture information and laboratory discoveries.

Cal Newport, in his book called *How to Win at College*, describes his method of studying as "the quiz and recall method." Newport states that "The details of this approach are simple. For any material you have to learn, be it complicated social science theories, foreign language verb conjugations, or physics equations, your basic study method should always involve quizzing and recalling. I don't mean this lightly. Asking yourself a handful of review questions doesn't count. You should build your entire approach to studying around the concept that making yourself recall specific information is the absolute best way to learn."[4]

For each unit of your anatomy study, I would encourage you to make that "specific information" an atlas plate or two, and then acquire wisdom layer by layer, over the course of your entire life, like tree rings of knowledge.

Quizzing and recalling. Cover those labels and name them. Look carefully at the structures they point to. On the particular plates that you think are important, do it until you never have to peek under that 3x5 card.

6. Forbes Method of Solidifying the Information in Your Brain: the best way to learn is to teach, especially at the cadaver table and in study groups.

Dr. Kamyab has good advice about study groups, as well. On the one hand, your cadaver table partners are a ready-made study group. On the other hand, a separate study group *per se* can expose you to contributions from five or six people *in addition to* your cadaver team.

In the chapter entitled "The Cadaver Laboratory," I will try to convince you that the best thing you can do *for yourself* in the Anatomy Lab is to come to the lab prepared. Let's just say you do that. Now when you get to the cadaver table, a fair number of your fellows may *not* be prepared. In that case, you get to do *my job* for a few minutes and teach the others. You have a captive audience, and when they realize that they're lagging behind you, they will become active learners.

I know what you're thinking: by teaching the others who don't know the material, you are rewarding their indolence. I absolutely agree with your revulsion at rewarding indolence. So you're going to have to be judicious in this. You might even want to put off sharing your wisdom until an after-hours session with those students who *do* prepare, and who *can teach you* a thing or two, in return. I just love those sessions, don't you? Everybody's in the teaching mode, and everybody benefits. *Docendo discimus*: by teaching we are learning. One of the biggest factors for success that I've identified is this: students who work with other students do better in my course. That's it.

Bradley R. Staats, in his excellent book *Never Stop Learning*, has many insights into what you're trying to do in professional school or graduate school. One of them is this:

"By working with the same team members, individuals found, they learned and dramatically improved their performance." He goes on to state: "The finding was as consistent as it was powerful: teams that had worked together before were dramatically more likely to deliver their projects on time, on budget, and with higher quality. Prior experience *together* was related to a 30 percent decrease in budget deviation and a 19 percent decrease in defects.[5]

This idea was the core of his doctoral dissertation at Harvard.

I can't speak for other disciplines, but this I know: loners do not do as well in Anatomy as group members do. Dissection table teams in the lab are excellent for this. So are study groups. So are "pick-up games," when several friends decide to go to the Lab together, on an ad hoc basis.

Here's a great analogy by Professor Staats: "[Let's say] you want a twelve-person jury to decide an individual's guilt or innocence. You have each member of the jury listen to one-twelfth of the trial. Would you then begin deliberations by taking a vote as to guilt or innocence? After hearing only one-twelfth of the trial, would you argue that you could decide the fate of the accused beyond a shadow of a doubt? Of course not. The members of the jury clearly would have very different information... To learn, you must gain insight into the eleven-twelfths of the situation that you haven't personally experienced. Remember the approach of the smartest person I know – ask yourself in every interaction what you can learn from the person sitting across the table from you."[6]

It's surprising that some students taking Anatomy class bring their bad habits from undergraduate experience to this study and expect to succeed. If you are thinking this way, you need to think again. If you skip the lectures (running Panopto at 2x instead), and if you do not come to lab prepared, and if you're not a member of a study group, you are handicapping yourself into a familiar pattern of failure. You may get by the imminent examination, but your knowledge will be superficial, and short-lived, and ultimately your patients will suffer. Yes, I'm aware that Tarzan was illiterate until he discovered a book in the jungle. He picked it up, and, all by himself, taught himself to read it. If you're Tarzan, you may succeed as a loner in your Anatomy class. If you're not Tarzan, you may not.

Students who struggle in my classes come to my office for counseling, and a high percentage of them answer negatively my first question: "Do you have opportunities to occasionally study with other students?" I think Anatomy in particular benefits when different individuals bounce ideas off from one another, quiz one another, challenge one another, hold one another accountable, and encourage one another.

You need to remind yourself that, while you're the most important person involved in your learning, you're not the *only* one involved. Your success, ultimately, depends on others almost as much as it depends on you. Therefore, you must select your mates from the clerisy, build relationships, inquire vigorously of others, and *take advantage of opportunities to teach other people.*

And when you meet with students in a group, be prepared. If everyone is prepared when you get together, you'll find yourself "drafting" on one another, like bicyclists do in a race: each member of the team periodically becomes the leader, and by following close in the leader's slipstream, all the members of the team benefit (even the leader).

Don't expect other members to do the advance work, and to teach you the material when your group meets. They will quickly figure out what you're doing, and they'll be meeting without you. As would I.

Throughout my long life, I have gone to national meetings of various societies, and there's always a big-shot plenary speaker after dinner. As impressive as his or her presentation might be, invariably I learn more *during* dinner, talking to the other anatomists, than I do from the speaker. This is because I bring my own questions and dilemmas to the dinner conversation, and, when you ask a question, you "carve a place" in your brain for the answer. I therefore work strategically in selecting my seat next to someone I respect and admire.

Someone significantly older than I. Someone whose responses to my questions can make me a better professor. I like to think that I contribute to those conversations, as well. In that way, the group at the banquet are an *ad hoc* study group, and I hope I've persuaded you that a study group is a great learning environment.

There is simply no better way to learn material well, than to teach it. Ask your professor about this: is it true that the best way to learn something is to teach it? I'm quite sure your professor will agree with me. The exchange of ideas is the *sine qua non* for learning. Whether you're a member of a banquet table group or a study group or a dissection table group, you'll be teaching each other. This is just what you will be doing throughout your professional life. That's what professional journals are all about. Even if you have just one other student to study with, working together can make a tremendous difference. Explaining to others and articulating difficult concepts is always time well spent.

Your group might want to try making up your own test questions, and presenting them to each other. You'll find out what every professor soon learns: making good, fair, relevant questions is more difficult that you'd think.

Several students recently told me that when they were learning the dermatomes, their group got together and drew those areas on the body of one of the members. They used "body paint," they said. Those students told me body paint is made to come off, but they found that it's not so easy. Commercial products for removing body paint didn't work very well. Since that time, I've learned that there are some home remedies to remove body paint, that involve cucumbers and milk, or baby oil. In any case, you may be wearing the paint for a while. That's a good thing; you'll never forget the fact that dermatome T-10 is on your belly button.

7. Forbes Method of Maintaining Your Attention: variety is the spice of Anatomy

The current trends in education affirm what Anatomy instructors have known for the millennia. If you were learning a new language, you'd be well advised to vary your exercises with reading, speaking, and learning vocabulary. If you concentrate on only one of these aspects, your knowledge will suffer, because of lack of variety. Musicians vary their repertoire for the same reason. Athletes, too, vary their workouts with speed work, strength training, and refining their skills. Everything from fartlek to muscle confusion theory is based on the idea of varying the protocols to enhance the psychological aspects of conditioning, to overcome plateaus in improving your fitness, and to avoid boredom.

Varying the venue for learning Anatomy is important, and this is built into most Anatomy courses, by virtue of the fact that Anatomy is studied in the laboratory, at home, and in lecture halls, in breakout rooms at school with other students, as well as (my favorite) the stacks in the library. Studying in different environments may not make beautiful music, but variety strengthens the anatomy student's abilities in the same way that variety strengthens the abilities of the musician.

In Anatomy, you'll find yourself varying the *ways you study*, as well. The idea of using plastic models, atlases, skeletons, on-line materials, your own body and *cadaver dissection* reinforces the information in a variety of ways.

If your study method is to learn one textbook, your knowledge will be inchoate and monochromatic. Rather, reinforce and expand the material by pondering it in several different contexts. Your dissection manual, for example, presents information in a concise, easy format. Attend the lectures in the course; I can tell you it's rare that students who "stream" the lectures do better than students who show up. Problem-solve in the anatomy lab by asking questions and doing difficult dissections. (This doesn't mean long, tedious sessions; it means challenging yourself as you dissect, knowing that the stakes are low if you make a mess of it.) Most of the time, students talk about the anatomy as they work in groups at the dissection table. Dissection of the cadaver is the ultimate in problem-based learning (PBL). Ponder your cadaver and compare it with the ideal anatomy found in the plastic models, atlases, and your own body (!)

It's what I call *learning in stereo:* it is a good idea to learn Anatomy in stereophonic sound. Here's the point: if you listen to music in mono, it can be very nice music, but it's not as interesting as if you listen to it in stereo or surround sound. The same is true for learning Anatomy: if you hear it (or read it, or watch it) in two (or more) different sources, it will be more interesting, and therefore more memorable, than if you learn all your Anatomy entirely from one book, or from one set of lecture notes.

I encourage you to get more than one Anatomy atlas, and more than one textbook. That second book serves you like the phrase, "In other words…" Most healthcare providers agree that getting a second opinion is almost always a good idea. You can buy used Anatomy atlases on Amazon for piggy-bank money. Buy a second, different, atlas (and a third one).

Let me give you some examples. Even if they're not required for your course, treat yourself to an old, inexpensive copy of *Gray's Anatomy for Students*, and an old atlas by a different author, and take a look at the structures you're studying in *Rohen's Color Atlas of*

Anatomy[7] (an atlas that presents photographs of skillfully dissected cadavers). These and other books are listed in the "Textbooks and Other Books" section of this book. I hope you have a chance to look at *Grant's Atlas of Anatomy*[8] (especially the 10th edition), which is not for the novice. It's one of my favorites, though, because I'm no longer a beginner. *Gray's Anatomy for Students, 4th edition*[9] and *Netter's Atlas of Human Anatomy 5th edition,*[10] are excellent for beginners. Learn your Anatomy in stereo, and you will remember it better. *Anatomy, A Regional Atlas of the Human Body*[11], by Carmine D. Clemente, is wonderful when used as reinforcement of what you're learning. If you aspire to a straight number (#1) for your academic place in the class, you should have a crooked number of Anatomy atlases. 'Nuff said.

Most Anatomy classes present the students with several different visits to the same anatomical information: anatomy students study for a quiz, then for the midterm exam, then for the final exam. They study once to prepare to dissect, and again in anticipation of my verbal quizzing at the dissection table, they study for the mocktical exam, and then they study again for the practical exam. They study the material (and forget it) over and over. But not to worry; forgetting is part of learning. If you re-learn it for "formative evaluations" as well as "summative evaluations," that information will stay with you into your dotage. Some people think that the harder it is to learn something, the harder it is to forget that something. (If that's true, then why can't I remember what I learned in Organic Chemistry?)

Many Anatomy courses incorporate some pathology in the curriculum, to give context to the study. The line between "normal" and "pathological" sometimes gets blurred: is it normal to have plaque in the lower end of the aorta? Well, it's common. In the 2000s, people live with compromises in their health that would have killed them a generation ago. Many of today's healthcare practitioners spend more of their time managing risk factors than treating maladies. Defining what's normal can be difficult: in Maine, every day the weather is abnormal, the road to Bangor is always worse than last year, and the biscuits never taste quite as good as usual.

One variety of study aids is leveraging the cadaver lab outside of scheduled lab sessions. The cadaver laboratory is the largest, most tech-driven and highest-yield study environment on your campus, and it is dedicated to one very specific activity, for one focused student group, for one specific course. And you have 24-7 access to it! You will be surprised how profitable an hour alone in the anatomy laboratory can be. And how enjoyable. It's quiet in there; there are virtually no distractions. Nobody's looking over your shoulder, telling you what to do, nor judging the quality of your dissection. And here's that old "learning in stereo" thing again: you have total freedom to examine other people's cadavers for a few minutes. (Don't dissect them.)

When time will allow, I like to watch Professor Marian Diamond's lectures on You Tube. *Integrative Biology 131: General Human Anatomy. Fall 2005*[12]. I think you will like them too. Marian Diamond's lectures in Anatomy are magnificent. She has 40 lectures on You Tube, covering every structure of the body. Don't be put off by the old-fashioned format: blackboard (greenboard, actually) and colored chalk, simple props, and the like. These lectures are comprehensive, engaging and wonderfully informative. If you're struggling in learning a certain area of the body, give the lovely and capable Dr. Diamond a try.

8. Forbes Method for Learning Nomenclature: break the word down, and look it up. Write what you've learned next to the term in your notes or in your anatomy atlas.

This principle has been discussed in detail in *"Chapter 3: Terminology, Etymology and Abbreviations,"* page 28, but, since it is an integral part of "The Forbes Method of Anatomy," it is revisited here.

You'll struggle with anatomical terminology if you learn the names of structures by rote. Rote is defined as "mechanical or habitual repetition of something to be learned."[13] Rote learning is the enemy of understanding. The better approach is to

1. break the word down into logical parts
2. hitch each part to something that will remind you of its meaning.

Early in your study, many of the structures you encounter will seem disjointed, and learning them can be arduous for the novice anatomist. At this point you will need to simply memorize the names of the structures. As your study continues, you will see how all these nerves, blood vessels, and muscles that you've learned, fit into the overall scheme of the body.

As you set about to learn the names of the structures, be on the lookout for opportunities to break each word down into component parts. The word "intercostal" indicates that these structures are between two costals. If you know that "costal" refers to the ribs, then you know exactly where to look for the structures bearing that name. Now consider where the "intercostobrachial nerve" is located. ("Brachial" means "arm.") Intercostobrachial nerve runs from between the ribs to the arm.

Let's take another example: when you learn the contents of the thorax, you will become acquainted with a little artery called pericardiacophrenic artery.

Let's break the word down into peri – cardiaco – phrenic.

1. peri: *Stedman's Medical Dictionary* defines this prefix as "around, about, or near."[14] Can we think of a word with these syllables in it? Maybe "periosteum" comes to mind. "Osteum" is bone, so periosteum is the membrane around, about or near the bone. How about periodontal membrane? "Dontal" refers to the teeth. It's the membrane that runs around, about or near a tooth, and holds it in the socket.

2. cardiaco: anyone who's known a person who's had cardiac arrest knows this part of the word refers to the heart.

3. phrenic: this might not be so familiar to you. "Phrenic" refers to the diaphragm in the same way that "renal" refers to the kidneys and "cardiac" refers to the heart.

So, when learning the name of this artery, your work will be easier if you consider that it's the artery that runs around, about or near the heart, and runs to the diaphragm.

The derivation of the word is the most important benefit of looking up a word. That's because later on you'll apply that derivation to other words with the same root, or the same prefix or suffix. Once you learn that "salpingo" means "trumpet," you will find yourself applying that root word to several structures in Anatomy: for examples, the salpingopharyngeus muscle in the pharynx, and the mesosalpinx, which is part of the mesentery of the pelvis.

I would like to disabuse you of the notion that you can get the same information from Wikipedia or Siri. *Get yourself a good, unabridged, medical dictionary. A late-model used one is almost as good as a new one.* You can buy a used 27th edition of *Stedman's Medical Dictionary* (the one I use) in "acceptable condition" on Amazon for ten bucks! When you're learning nomenclature, always keep a good medical dictionary close at hand. It will give you the derivation of each part of the word, and make it easier to establish the word into your brain.

Bottom line: you'll remember a new word better if you look up the definition and determine its derivation.

9. Forbes Method of Taking a Test: take it after a good night's sleep.

When I was doing undergraduate study, I found it helpful on some occasions to "pull an all-nighter," in order to overcome my prior procrastination. The next day, after the examination, there was plenty of time to sleep.

When you're involved in a professional program with a rigorous Anatomy course, that "plenty of time to sleep" is usually a mirage. Instead of sleeping after the exam, there'll be new material to learn that next day, and the next. In almost all instances, it's silly to pull an all-nighter studying in professional school or graduate school. It's a well-known formula leading to anxiety and underperformance. You can't drink enough coffee to compensate for the resulting lack of keenness. Staying up all night and sleeping the next day is akin to the Synchronized Swimming coach advising her swimmers, "There's plenty of time for *breathing*, after we leave the pool!"

Most people do not have much luck cramming Anatomy. My roommate in undergraduate college, a psychology major, was a brilliant kid, and he would stay up all night before his exams reading psychology, and he achieved a very fancy GPA. I remember him telling me, "Bill, *of course* I can recall the material when I'm taking the test. *I just read it!*" His approach would not have availed him in Anatomy. Psychology is concepts and theories; anatomy is relationships and many facts.

It's ironic: cutting back on sleep will be detrimental to your academic performance, and yet it's often the first step people take when faced with academic difficulty.

Get a good night's sleep the night before your exam. *I know everyone tells you that, but you feel like you're different. You feel like you can stay in screech mode all night, and still be keen for the exam.* You aren't, and you can't. You need your rest. Even GOD rested! In the book of Genesis, it records that after working hard for six days creating the world, God took a rest (Genesis 2:3). You should too.

10. Forbes Method of Practical Anatomy: If you are confronted with a structure you can't identify, find something that you *can identify*, and work from there.

In my time spent in the Anatomy Lab, I'm often called over to a dissection table, where a student will point out a structure on a model or a cadaver, and say, "What's that?"

Now, I must confide in you: a large percentage of the time, I have no idea what that is. Here's my approach to this familiar dilemma: I find a structure that I *can* identify, and I work from there. Here's an example: when dissecting the gluteal region, after you've reflected the gluteus maximus muscle, you will see half a dozen muscles, many nerves and neurovascular bundles, some tendons and bones (*Netter 5th ed.* 490, entitled "Arteries and Nerves of Thigh: Posterior View"). Some smart-alec fellow student will point out a structure in that gallimaufry and ask you, "What's that?"

Where do I begin? My approach is to *find something I know*; in this case it would be the very prominent, obvious sciatic nerve.

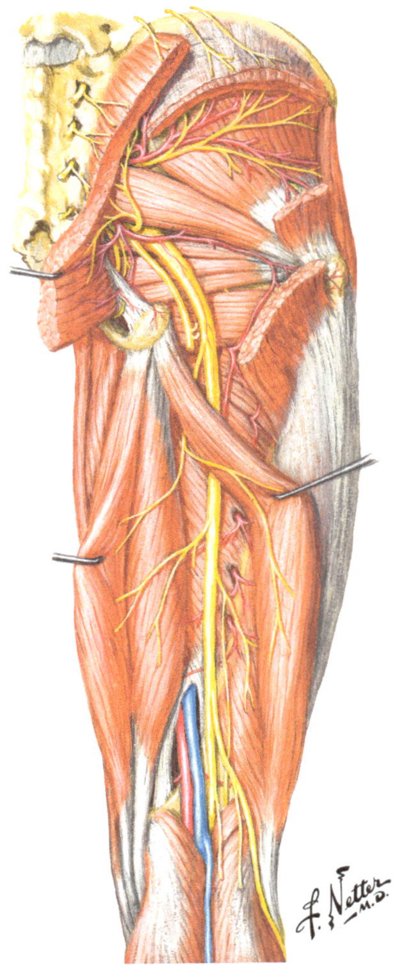

Netter Atlas of Human Anatomy, 5th Edition 490. Copyright 2011 Elsevier Inc All rights reserved. www.netterimages.com 1960

It's almost impossible to miss, because it's the biggest nerve you've ever seen. I'm reminded that the sciatic nerve emerges just below the piriformis muscle. O.K., there *that* is. Once piriformis is identified, I can confidently state that the neurovascular bundle emerging just superior to it is the superior gluteal artery, vein and nerve. I know that those particular structures run in the fascia between gluteus medius and gluteus minimus muscles, allowing me to distinguish one of those muscles from the other. And so it goes.

When you're stumped in your dissection, find something you do know, open up the relevant page of the atlas, and move from one structure to another.

11. Forbes "Tricks" in Learning Anatomy: use mnemonics and the like

This topic has previously been discussed in *Chapter 4: Acquiring Knowledge*, page 56. Since it is an essential and virtually universal technique to master Anatomy, it deserves a place here in *Chapter 6: The Forbes Method of Anatomy*, as well.

To learn Anatomy, there are several approaches that I call "tricks," that are perfectly legitimate to use. They can make it easier to memorize the names of a large number of structures, their function and the physical relationships between them. As described above, you should learn the material in surround sound, by utilizing books, atlases, plastic models and plastinations, bones and skeletons, white-board drawings, study group discussions, lectures and symposia, on-line study aids, and the ultimate study tool: the cadaver. And in addition, there are helpful mnemonics to buttress your study.

The use of mnemonics is common and popular among most students of Anatomy, and has been for the millennia. The most common type of mnemonic is a sentence that prompts us with the words in some list we need to memorize. Here's an example: branches of external carotid artery, listed in order, can be brought to mind with the mnemonic "Some anatomists like freaking out poor medical students."

some: superior thyroid artery
anatomists: ascending pharyngeal artery
like: lingual artery
freaking: facial artery
out: occipital artery
poor: posterior auricular artery
medical: maxillary artery
students: superficial temporal artery

The use of mnemonics can be helpful, but, like a twelve-year-old with makeup, it can be overdone. Many students catch the "mnemonic plague" when studying Anatomy. My experience has been that students often remember the mnemonic, but forget what each part stands for, or forget where to begin. For those students, mnemonics represent a "crutch" for learning the material, rather than wisdom and understanding. Bear in mind that the wisdom and understanding will be needed at your next steps: the *application* of the Anatomy to the disciplines of Physiology, Pathology, and others.

You therefore should develop *wisdom* about Anatomy. Here's an example: which nerves innervate the individual muscles that move the eyeball? There are at least two ways to commit this information to memory. Each is equally acceptable.

1. Perhaps the most common way is to memorize the *mnemonic* "LR_6SO_4Rest by 3." (Some people write it LR_6SO_43.) **L**ateral **r**ectus muscle is innervated by cranial nerve 6 (VI), **s**uperior **o**blique muscle is innervated by cranial nerve 4 (IV), and the

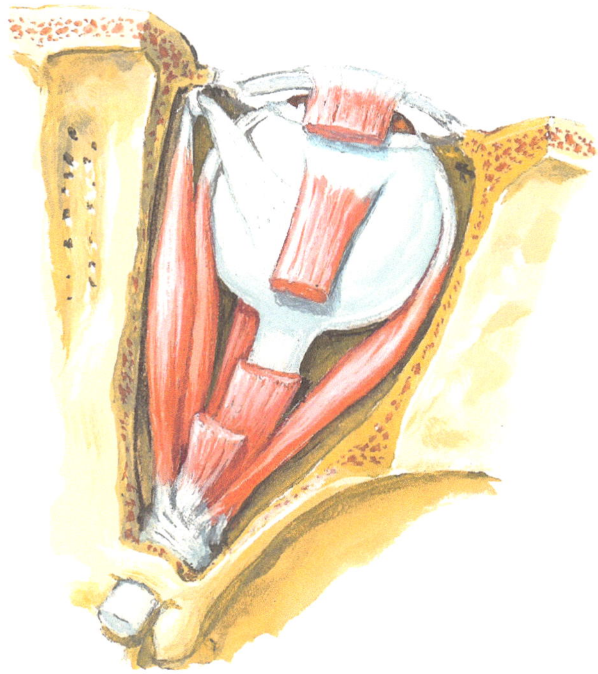

Netter Atlas of Human Anatomy, 5th Edition 84.

www.netterimages.com 3521

rest of the muscles that move the eye are innervated by cranial nerve number 3 (III). LR_6SO_4Rest by 3. Got it.

2. My preference is to apply the information a little more directly. To wit:

Think about lateral rectus muscle, in the orbit. What can we conclude from lateral rectus muscle's name? It tells us that the muscle pulls the eyeball in a *straight* (rectus) direction from the *lateral side*.

Look at the picture above: what happens when lateral rectus shortens in length? (Remember: muscles always *pull*; they never *push.)* Being attached to the lateral side, when it shortens in length, it rotates the eyeball in a lateral direction, *taking it away* from the midline of the body.

What do we call it when someone takes someone else away? They *abduct that person.* Lateral rectus muscle abducts the eyeball. So it makes sense that the muscle abducting the eyeball is innervated by a nerve called "abducens" nerve, the sixth cranial nerve, CN VI.

Similarly, superior oblique muscle. Can you see it in the drawing above? How does this muscle approach the eyeball *obliquely?* By passing through a pulley, and then turning back to the eyeball. Dr. Netter has drawn it beautifully, above (*Netter 5th ed.* 84, "Extrinsic Eye Muscles" above).

The word for "pulley" in Anatomy is "trochlea." So the nerve innervating this very unique and interesting muscle is called "trochlear nerve," (CN IV).

So *abducens* nerve innervates the muscle that *abducts* they eyeball: lateral rectus muscle. *Trochlear* nerve innervates the muscle that goes through the *trochlea* (the pulley): superior oblique muscle. All the rest of the muscles of the eyeball are innervated by cranial nerve III: oculomotor nerve.

Given the two options above for learning innervation of muscles in the orbit, my personal preference is #2, but that's just me. Either way is acceptable. You should think about a mnemonic construction in the same way that you think about your mother-in-law: you want to have a good relationship with her, but you don't want her to boss you around.

Here are some other "tricks" for remembering information about structures:

Applying clinical information to the structures tends to firm them up in the learner's mind. Example: if you pound your "funny bone" on the table top, the tingling sensation will tell you the sensory distribution of the ulnar nerve. Another example: studying the parts of your pharynx involved in the gag reflex, (with a long-handled spoon, say), is a memorable way to learn some of the sensory distribution of the "gag nerve": glossopharyngeal nerve.

A *joke, a play on words or an interesting fact* about the structure can help remember it. I often use the mastoid process as an example. You can see mastoid process in any anatomy atlas; in your *Netter Atlas 5th edition* it's plate 6. To early anatomists, mastoid process ("mast" = "breast," "oid" = "resembling") resembled a breast. In an attempt at humor, I tell my students, "If that bulge on the bone makes you think of a breast, I'm worried about you. Those early anatomists were mostly male, and apparently they didn't get out much." But remembering those anatomists' questionable preoccupations can help solidify the fact that those interesting bumps really are *mast-oid*. You may never feel the same way about that bony bump behind your ear.

One more idea: it helps to *learn a marvel* (there are many) about this miraculous structure called the human body, that we are privileged to study. An example of one of those marvels: your friend calls your name as you cross the campus. You know immediately from which direction the sound is coming. How do you know? How can you determine from which direction a sound is coming?

That sound strikes your left ear at a different time than it strikes your right ear, because one ear is farther away from the source. The difference in those times can be as little as 0.00003 seconds.[15] From your ear, the sense of hearing travels over the vestibulocochlear nerve (CN VIII). Each ear transmits its sensation over vestibulocochlear nerve to your brain, which interprets the difference in time, and processes it to tell you that your friend is over... *there!* Marvelous! If the fact that your brain can distinguish sounds 0.00003 seconds apart doesn't amaze you, then I'm worried about *you*.

You may have noticed your dog "cocking his head to one side," when trying to understand your commands. Some people think it's an attempt to comprehend better what you're saying, by looking and listening from a different position, which places its ears in a different configuration relative to incoming sounds.

Vestibulocochlear nerve brings the ability to hear sounds. That's a miraculous feat. Even more miraculous is the ability of the *two* vestibulocochlear nerves to team up and localize the direction of those sounds.

12. Forbes Method for Groups of Facts: rules and exceptions

To learn a group of facts, The Forbes Method for Groups of Facts organizes them into *rules*, and *exceptions*. For example, here's a *rule*: all the "juice" that flows from your head is stimulated by one cranial nerve: facial nerve (cranial nerve VII). Period. This "juice" that facial nerve stimulates includes saliva, runny nose, tears.

That's the rule. It will allow you to answer 95% of questions about the innervation that stimulates production of the various fluids in the head.

However, there are two *exceptions* to this rule: parotid gland is stimulated to make *its* juice (saliva) by glossopharyngeal nerve (cranial nerve IX). The other exception is the production of sweat, which is stimulated by the sympathetic nervous system.

So here's a rule to live by in Anatomy, bearing in mind that there are two exceptions to this rule: "Juice to the head? Facial nerve. Any exceptions? Yes: two."

I think it is always good advice to approach Anatomy in terms of rules, and exceptions to those rules. After that, with the rules firmly in your mind, it's easy to add details to those rules.

Don't try to learn all the details of Anatomy in your first pass through the reading, whether it is your notes or a textbook. Take several passes through. The first time you go over the material, try to identify broad principles, for example: the seventh cranial nerve stimulates almost all the "juice" produced in the head.

Then in subsequent passes through the material, try to hang additional facts on the broad principles. Those additional facts may be: exceptions to the rule or details supporting the rule.

Here is an example of details supporting the rule:

Rule: juice to the head? Facial nerve. There are two exceptions to that rule. Now let's take it a step further: let's think about that facial nerve, which is cranial nerve VII:

Above the level of the teeth, all the "juice" is stimulated by one branch of the seventh cranial nerve: greater petrosal branch.

Below the level of the teeth, the "juice" is stimulated by another branch of the seventh cranial nerve: chorda tympani branch.

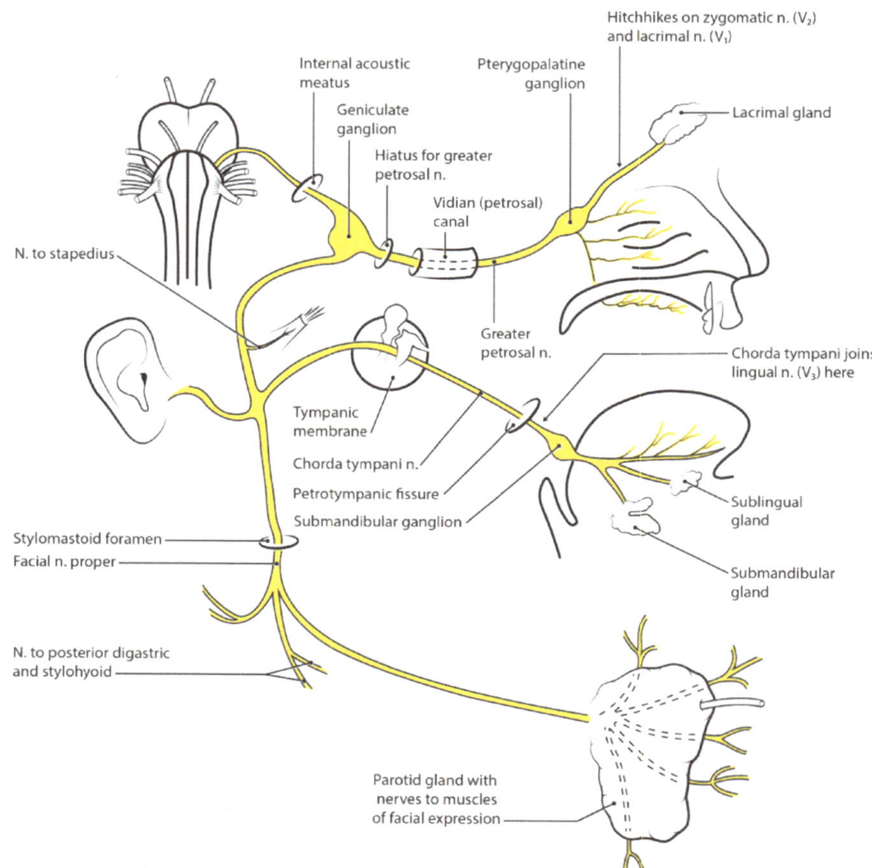

Facial nerve (VII)- WCF

A great deal of your work in Anatomy can be approached in this way. It's a way of organizing the material that makes it easier to learn. Your instructors have rules, exceptions and details that they've developed, and they'll tell them to you as you go along.

Here's another example: how does sympathetic innervation reach the myriad structures throughout the head?

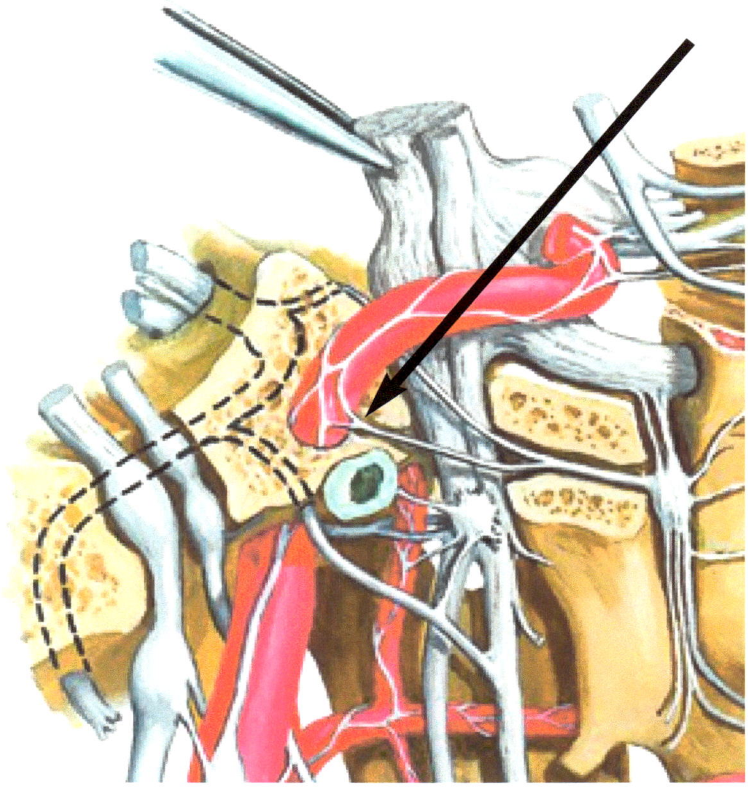

Netter Atlas of Human Anatomy, 5th Edition 130.

Here's the rule: it does so by forming a plexus that follows ("hitchhikes on") the carotid arteries and their branches. Can you see them in the image above? Since the carotid arteries and their branches run pretty much throughout the head, the hitchhiking sympathetic innervation enjoys the same distribution.

Any exceptions? Yes, one: deep petrosal nerve. You can see the nerve plexuses, glommed onto the arteries, in the drawing above, *Netter 5th ed.* 130, entitled "Autonomic Nerves in Head." You can also see the deep petrosal nerve, beautifully drawn, and pointed out by the black arrow. There's a rule, and the only exception to that rule.

Here are more examples of rules and exceptions:

Seven Rules and Exceptions for Head and Neck Muscle Innervation

1. All of the *tongue muscles* are innervated by hypoglossal nerve, ***except*** palatoglossus (vagus nerve).
2. All of the *palate muscles* are innervated by vagus nerve, ***except*** tensor veli palatini (trigeminal nerve).

3. All of the *pharyngeal muscles* are innervated by vagus nerve, ***except*** stylopharyngeus (glossopharyngeal nerve).
4. All of the *laryngeal muscles* are innervated by recurrent laryngeal nerve ***except*** cricothyroid (external laryngeal nerve).
5. All of the *muscles of mastication* are innervated by mandibular division or trigeminal nerve.
6. All of the *infrahyoid muscles* are innervated by cervical plexus, specifically via ansa cervicalis.
7. All of the *muscles of facial expressions* are innervated by facial nerve.

There. Rules and exceptions comprise the Forbes Method for Groups of Facts. Thinking in terms of rules and exceptions will serve you well, as you pursue your study of this amazing body.

Conclusion

I'm sure you will not adopt all the strategies listed above. However, you may want to select one or several to add to your armamentarium in accomplishing the important work you're setting out to do.

CHAPTER 6A: Discourse with the Donor

She came to my bedside, without the sound of footsteps. Nonetheless, her presence was real, in the same way that mist, or fog, is real.

"It's late," I said. "Could you come back some other time? I'm exhausted from that three-hour lab we had yesterday." Sleep had come quickly, but not in sufficient quantity. At 3:00 AM, I wasn't in complete control of my mouth, or my tongue inside it.

"I know you're exhausted. You try too hard. You jump from table to table like a hungry mouse in a maze. And when you do, your dissections border on mutilation. You're trying to be all things to all people." A picture began to form in my mind. It was a crazy man running hither and yon, mumbling garbled speech and sweating profusely. Then I noticed she was still talking. "You're doing something I call 'replacing cerebration with celerity.' You can't do a good job like that."

"I know. You're right, I'm afraid." I thought about the reasons for working like that. "But I'm in a hurry," I said. "I have too much teaching to do, and not enough time to do it. During a laboratory session, there are lots of students who need help. I can't afford the time required to do fastidious dissections."

"And you're casual in your barrier techniques. Yesterday you poked my skin with your finger, sans gloves! Other times I remember you've jabbed something with your pen and then put it in your shirt pocket. This is not good. Next you'll be poking my pancreas with your fork."

"I've got things to do. It takes time to glove up continuously. When there's a hint of mold in the lab we're required to change gloves every time we move from table to table. That's a big bother, and it takes time that I simply don't have. Do you realize that when I was in school, nobody wore gloves, ever?"

"What I realize is that you're in a hurry. It's obvious to me, and to your students, as well. My! How I crave the velvet touch of those orthopedic surgeons who came to the lab and dissected my knees."

"That was during a Clinical Applications session," I said. "These students are learning anatomical relationships, and the clinical application of those relationships inspires the students. However, watching orthopods do surgery on dead people is not what the students are here for."

"Well, I wish I'd had that knee replacement surgery done while I was living, and not a year after I died. Did you know that some surgeons can tie a knot with just one hand? That guy's hands glow in the dark."

"Darling," I responded, "Tying a knot with one hand just takes practice. It's difficult, but it can be learned, like learning to fold a fitted sheet. And as for dissecting, you've seen me doing prosections alone in the Lab on Saturday mornings; I make Nobel-prize-winning dissections. But when I have six tables of students to attend, and when the skin needs to be taken off, I just, well… I take the skin off."

"And, along with my skin, you took my great saphenous vein, and my beautiful muscles of facial expression, leaving my body looking like I'd slept in it."

"I'll try to do better," I said. Her criticism was beginning to wear on me, and my fondest hope was that she would leave.

She leaned forward and, almost affectionately, said, "I miss you when you don't stop at my table, Habibi. You're The Man, and you're a good one." She nodded in agreement with her own comment.

Then, apparently she realized what she'd said, and she stopped. She saw the surprised pleasure on my face. At her stage of dissection, she lacked the ability to blush. She smiled as demure a smile as someone can who's missing one of her risorius muscles. She thought for a moment, apparently decided her comment was too complimentary, and immediately took some of it back. She said, "And your singing in the Lab on Saturday morning sounded like a rusty hinge. It was so morbid that it made me sad."

I responded without thinking. "You, of all people. Here you are, complaining about morbidity! You can go now. Just disappear, like the ghost of Christmas past."

She stood, contemplating the polish on her few remaining fingernails. It was clear that she had no intention of leaving. She glanced casually around the room and saw my notes and papers. "Whacha doing over there?"

"If you must know, I'm writing a book." I anticipated her next question. "It's about how to learn Anatomy – that's the only thing I know anything about."

"Am I in it?" she said, thoughtfully riffling through the pages.

"Oh, you're in it all right. You're the main character – the Alice in my Wonderland."

"The Alice in your Wonderland. Huh," she said to herself thoughtfully, while she was preoccupied with the pages of my notebook. Sliding her denuded hand over my desk, she knocked a pile of pages to the floor.

I was indignant. "All right. That's enough. I've just about had it with this. I'm getting up now."

"If you get up, your dream will be over, and I'll be gone." She kept reading.

"Oh, I'm getting up, all right. And I'm warning you: I'm naked, and there are aspects of morning arousal that can be embarrassing. It's not sexual, of course. But at my age, nobody wants to see that…"

I looked around the bedroom. There was no sign of her. All that remained was the faint, pleasant aroma of formaldehyde.

CHAPTER 7:
The Cadaver Laboratory

The Ultimate Learning Tool

When you visit my anatomy laboratory during a busy dissection session, the first thing you should do is to listen. That sound you hear is the voices of students teaching students. Can you hear it? Isn't it lovely? It's nothing less than beautiful music. Students are gathered around the table like a string quintet, and they're practicing the same way the quintet practices, working together harmoniously.

Studying Anatomy in the cadaver laboratory is the embodiment of all the compelling new instruction methodologies that they teach in those dreary in-service lectures that we professors are required to attend. Turns out that students working in my laboratory are exhibiting many of the principles found in cutting edge pedagogy and the educational journal articles that result. The exercises the students are completing demonstrate the best pedagogical aspirations of professors in many other disciplines. Teams dissecting in the Anatomy laboratory are doing PBL (problem-based learning), TBL (team-based learning), teamwork, problem solving, and student instruction and student research all wrapped into one experience. Furthermore, the cadaver-dissecting anatomy laboratory has been demonstrating these "new" educational concepts for generations.

There is no need to defend the benefits of the cadaver dissection laboratory. Anyone who approaches the issue without the imposition of other factors (finances, time, space) will conclude that dissecting is the best way to learn the anatomy of the human body (or any other body). Let me put it this way: occasionally a Gross Anatomy professor gets away from professing for a year or more. Maybe there's a family emergency, or a prolonged sickness, or maybe he tries to retire and it doesn't work for him. Do you know what he wants when he returns to the class? He wants a cadaver to dissect. He knows that dissecting a cadaver is the very best way to learn (or, in his case, to *re-learn*) Anatomy.

Most important of all, the laboratory sessions are composed of *students teaching students*, and most educators will agree that *teaching* is the best way to learn. "See one, do one, teach one" is a time-honored adage, and it's never more true than in the Anatomy Lab. There's a reason why in most labs, one member of each team is assigned a title like "reader," or "leader," for the session. There is real value in teaching the material. Frank Oppenheimer, the famous American particle physicist, was famous for his witty observations, including "The best way to learn is to teach."

In my career I've mostly worked with students pursuing healthcare professions, but not all the dissectors in a cadaver lab are necessarily so. I've seen cadaver dissection teams composed of biomedical engineering students, I've loved working with graduate students putting together a master's degree, or a Ph.D., and I've gotten to know teams of students studying the challenging and fascinating major called Medical Illustration. Each discipline has its own priorities and emphases, preparing students for a broad range of worthy careers. And they come together, in the cadaver laboratory.

At some schools, the Anatomy Laboratory becomes a sort of "clubhouse" for the students of Anatomy, whether they are professional students, graduate students, teaching assistants, or undergraduate students. To be excellent in a Gross Anatomy course that uses cadavers, students must hie themselves to the laboratory outside the regularly scheduled laboratory sessions. That means lunch hours, evenings and/or weekends in most cases. When they do come to the lab after hours, those students or teaching assistants will tend to encounter the same comrades in the lab, over and over. The lab, for them, becomes a "home away from home." Those anatomically-inclined students develop a strong camaraderie, and they stick together. They help one another, and sometimes they even fall in love. Yes! It has happened, *more than once*, in my Anatomy laboratory.

The "down side," I suppose, is that there is no eating nor drinking in the Anatomy Lab. Not even coffee. No picture-taking is allowed, and no photography of any kind, unless it's expressly permitted by a faculty member.

The Early Anatomists

Early anatomists, of course, commonly dissected cadavers of their own age. For example, in the 19th century, when a human being's life expectancy was half of what it is today, most anatomists were dissecting their peers. Today, in my laboratory, we find that most of our cadavers died at age 70 years old, or older. For the twenty-somethings and thirty-somethings who work and dissect in my classes, that seems very elderly. For some of us professors, those cadavers are *our* peers.

Early anatomists were also dealing with the need to work secretly, and quickly. In many cases, dissection was illegal. Furthermore, they worked hurriedly, because the body decayed so rapidly. Nowadays dissecting is completely legal as long as it takes place within licensed institutions, and in most cases the bodies are embalmed. If they're kept from drying out, the embalming solution can preserve those bodies for many months.

In the United States, dissection of human bodies was illegal until the mid-nineteenth century. In fact, if you Google "William Forbes, anatomist," you'll find an account of a person I hope is a relative of mine: William Smith Forbes, Demonstrator of Anatomy at Jefferson Medical College in the 1880s. He evidently had the same appreciation for the laws restricting dissection that I presently hold for laws restricting speed on the highway. In 1882, he was indicted on eight counts by a grand jury for "complicity with resurrectionists in the despoliation of graves in Lebanon Cemetery."[1]

It seems that late at night on Dec. 4, 1882, in Philadelphia, Pennsylvania, the superintendent of the Lebanon Cemetery, together with three accomplices, was arrested as they were furtively digging up a corpse in the cemetery. The next day they were arraigned and charged with the crime of grave-robbing. People in those days had such strong feelings about the crime of grave-robbing that a mob assembled and threatened to lynch "the graveyard ghouls." The police investigation revealed that the scofflaws under arrest were actually doing so under the direction of the esteemed Dr. [George] William Smith Forbes. He was arrested on or around Dec. 15, 1882, and his bail was set at $5,000.

Three months later, Dr. Forbes stood trial for the crime of conspiracy to commit grave-robbing. Dr. Forbes explained to the court that many medical students studying gross anatomy were unable to access cadavers to dissect, because of strict laws in Philadelphia made it impossible to obtain enough cadavers. This situation prompted some professors to hire professional grave robbers, called "resurrectionists," to provide bodies to medical schools.

Forbes was acquitted of conspiracy, and the unpleasant controversy had a salutary effect. The Pennsylvania legislature passed a new law, which incorporated amendments written by Dr. William Forbes, in June 1883. The law instituted a state board, charged with regulation of cadavers used for academic purposes. Within four years, Forbes was promoted to Professor of Anatomy and placed in charge of the General Surgical Clinic at Jefferson College. There he continued to teach anatomy and surgery for more than fifteen years.

So this person, who I hope is my great-great-great-uncle or something, was a grave robber, but for a good cause: he robbed graves in order to dissect and learn the human body.

Lay People in the Lab

I find that there's great interest in the cadaver lab by lay people. They seem intrigued by the idea that I spend a large part of each day taking people's bodies apart. There is a great deal in the lay literature about anatomy labs; I personally own at least five books about the anatomy lab experience, written by people who were not anatomists. These people went through a semester attending anatomy lab with students, and then wrote about the experience. Some of them are listed in the "Textbooks & Other Books" chapter of this book, under "Books by Laymen About the Cadaver Lab Experience." If you're interested in the cadaver laboratory experience, they are worth reading.

I myself have been milling around Anatomy laboratories since before you were born, like a "rink rat" at an ice hockey arena, and I still get excited at the prospect, each time I enter the lab. I never leave the lab without the feeling that I've gained something in that day's experience. A bit of knowledge, a relationship between structures, a clinical application, new information promulgated by a fellow professor, a mnemonic taught to me by a student, something. Always. You will, too.

Your First Patient

It's common for people who work in Gross Anatomy to say, "The cadaver is your first patient." I suppose that's correct, although the differences are obvious. We should all be grateful that our surgeon's first cut on human tissue was on a dead person and not a living one. I like to say that the cadaver is the best Gross Anatomy *teacher*. The cadaver will teach you a great deal, if you take care, and pay attention.

Indeed, many Anatomy professors explain to their classes that while their first "patient" is deceased, and therefore beyond help, the dead donor body on the dissection table will "help" the student learn a great deal about the healing arts. For that reason, and others, we treat the "patient" with respect – similar to the respect the students will show for a living patient – and we must maintain confidentiality as well. We can't betray confidences outside of the class about cause of death, tattoos, and the like. Their living relatives rightfully expect and deserve this confidentiality, and cadaver laboratories require it.

Isn't it interesting that I've worked in Southeast Michigan for thirty years and I've never even *heard of* a person identifying a donor body in an Anatomy laboratory? Never. It simply doesn't happen.

It pains me to see students in the cadaver lab who don't appreciate this once-in-a-lifetime experience. It reminds me of the old saw: education is the one commodity where people try to get as little as they can for their money. Is that true of you? Are you one of those people who comes up to me after class and asks a question which is essentially, "How little do I need to know to pass the test?" If you are one of those, you are making a big mistake. Dissecting and learning the cadaver is an experience that likely won't come again, and it's a serious error in judgement to miss this opportunity by holding back your maximum effort.

The pleasure of dissecting, according to Bill Hayes, in his excellent book *The Anatomist*, "…lies in making order of the disorder, in tidying up what looks messy. It is an art well suited to fastidious types such as myself. 'Make it pretty,' [the laboratory instructors] often instruct, but it's true – that is exactly what one strives to do. When done well, dissecting is very pleasing aesthetically."[2]

Later in his book, Hayes re-thinks the subject, and states, "That the body is structured in such a consistent, organized way is another reason I find dissecting supremely satisfying... In fact, perhaps I misstated it earlier. Dissecting really has nothing to do with making things orderly. The order is already there, just under the surface. The anatomist only has to uncover it."[3]

Believe it or not, dissecting can be one of the most satisfying things you can do. Very often it is done all by your lonesome, in the laboratory, where the solitude tends to focus one's attention on the delicate tasks at hand. It's like working on a difficult jigsaw puzzle: you work for an hour or so, and when you stand back and look, you're surprised at how much you've accomplished.

And you should resolve to do that: stand back and look. You've memorized the Anatomy even before you dissected the cadaver, but once the dissection is completed, *then* you can begin to learn from this magnificent study aid that you've created. Go through each structure like a kid learning his ABCs: repeat it aloud for someone else. Act like your tablemate had just arrived in the Laboratory: begin at the beginning, and go systematically through it. Remember Forbes' Method of Reinforcing the Information: The best way to learn is to teach.

Then go to other cadavers, the ones that other groups have dissected, and go through their cadaver in the same way. You'll encounter The Paradox of the Cadaver: each cadaver is the same as the others, and, at the same time, each cadaver is unique.

When you're finished, be sure to scrupulously wrap the other group's cadaver, and zip up the body bag. This attention to detail is common courtesy, in order to preserve their excellent work, and I'm sure they'll do the same for you.

Privacy

It would be a mistake to make the cadaver laboratory visible to passers-by or gawkers. Therefore, the lab is locked, and only the participants in the class are given access to it. We don't prop the doors open. We don't allow children in. (At our school the restricted age is under 15.) We keep our voices down. All information that might provide identification of the donors is strictly confidential.

First Aid

All Anatomy laboratories have First Aid kits, for treating minor accidents that occur. Any cuts should be washed completely in the sink, treated with Neosporin or other antiseptic, before applying a Band-Aid or bandage. Bear in mind that the embalming solution kills all pathogens; you're not likely to contract an infection from your cadaver.

All laboratories include an eyewash station, for your use if you accidentally splash your eyes with cadaveric fluids or tissues. You should become familiar with their use *before* you have an accident requiring their use.

The Cadaver Tables

The tables the donor bodies are lying on have traditionally been plain stainless-steel tables with a drain at one end leading into a bucket below. Nowadays we have more sophisticated tables, some with a "down-draft" feature, which means the tabletop is perforated, and a vacuum system is attached beneath. The vacuum takes the formaldehyde smell down through the tabletop, and directly out of the lab. These systems work well, but there is a steep learning curve about regulating the time and pressure of the downdraft. If there's too much pressure or if they're left on too much of the time, the cadavers dry out. If they're not left on enough, or if the draft is weak, the system is ineffective. Once the proper levels of draft are figured out and adopted, there are generally no problems with these systems.

There are also available stainless steel "clamshells" that cover the cadavers when they're not being studied. When the clamshells are closed, they form a stainless steel cover in the shape of a tent over the body. They can be opened on each side like a clamshell, and fastened open while the cadaver is being dissected or studied.

The clamshells are no substitute for the body bags, however. The cadaver should be placed in a body bag, and then the clamshell covers the bag. The body bags are zippered: the zipper either swoops down one side, or courses straight down the middle of the top of the bag. Almost always they are white, which should help in identifying instruments left inside, prompting you to remove them scrupulously before closing the bag.

Most morgues will store the bodies at about 38 degrees Fahrenheit, which is cold enough to stifle growth of most microorganisms, but warm enough not to freeze the tissues. Cadaver labs are kept at room temperature.

On balance, these recent improvements greatly enhance the cadaver laboratory. They cut down on the desiccation of the cadaver and the smell in the lab. Since formaldehyde in prolonged or large doses is a carcinogen, that's always a positive development.

Tissue Receptacles

There are two key rules about the human tissue receptacles in my laboratory:

1. Never put anything but human tissue in those receptacles.
2. Never, *ever* put anything but human tissue in those receptacles.

One of the worst offenses you can commit in my Anatomy Lab is to put paper towels or gloves into the human tissue receptacles. These receptacles, tissue bins or tissue bags, are colored red, and they are the repository for nondescript, unidentified scraps of human tissue, and they are labeled accordingly. They should receive human tissue *only*. Don't put paper or any other non-human material into those bins! The folks at the crematorium won't be happy, so your instructors won't be happy, so *you won't be happy*, if you commit this heinous act.

Ersatz Cadaver Laboratories: Plastic Models, Prosections, Plastinated Specimens, Virtual Dissection, Animal Dissection

Given all that's beneficial about cadaver labs in professional education, it's surprising to find the cockroach in the caviar: some parsimonious professional schools are gradually doing away with cadaver labs, and moving toward labs that use exclusively plastic models, prosections, plastinated specimens, computer virtual dissection programs, and the like. You should be aware that their motivation for moving to these alternatives is *not* to improve their students' education. *Their motivation for doing this is to save money.* In a breathtakingly sophistic statement, a particular school administrator informed me that the school was replacing the cadaver lab with a course using only plastinated specimens. She said, "No longer will the students have to waste their time dissecting. They can get what they need from plastinated specimens, and then go to the library to actually learn." I offered my objections, and vigorously promoted the traditional practice of studying human cadavers. In infuriated Machiavellian fricatives, she said, "Forget it, and back off!" This is naiveté in deep disguise. There's more to eating than shoving food down one's gullet, and there's more to learning human anatomy than memorizing nonsense syllables in the library.

In most professional school or graduate school anatomy courses, the students have honed their intellectual appetites until they've created in their minds the demand for cadaver dissection. This has prompted most professional schools to offer such a class to their students. A prospective student should carefully explore the curriculum of his or her intended professional school. If a talented young person attends a professional school where students are not required to actually dissect a donor body, that person may later regret missing this once-in-a-lifetime opportunity. Evaluating the value of the cadaver lab is an example of the dilemma that you're going to wrestle with for your entire healthcare career: the tension between the doctrines of the authorities that you choose over your life, and your own personal persuasions.

Professors of Gross Anatomy, who are the people who should know, periodically find it necessary to defend that pearl of Anatomy pedagogy, the cadaver lab, against assaults by the administration, and even from their department chairs. Every few years, in a Sisyphean redux, Anatomy professors are compelled to make the case that the only way to learn Anatomy for clinical practice is by cadaver dissection.

While you're working with your donor body, keep in mind one important benefit to cadaver dissection that cannot be equaled in other forms of anatomy instruction. "Surface anatomy" becomes an important part of our study in the Anatomy Laboratory. Health practitioners commonly examine the outside (exterior) aspects of their patients' bodies, and in doing so, they draw conclusions about what's going on beneath the surface. Several books take this approach; my favorite is *Color Atlas of Surface Anatomy*, by Backhouse and Hutchings. It gives valuable information about Gross Anatomy, and provides clues about what your clinicians are doing when they are listening, palpating and performing auscultation.

Some schools specify that their students do only virtual dissections, on a laptop computer. Some schools do prosections only, meaning someone has already dissected the cadavers; the students, reminiscent of a children's scavenger hunt, merely look at the structures that have been dissected by someone else. Some schools are doing even more creative things to try to cut corners: starting with a skeleton, the students are required to *add, with clay*, the muscles, the blood vessels and nerves, and build their own "cadavers." If you are a professional school administrator and you begin to entertain notions of taking the cadaver experience out of your Anatomy course offerings, I suggest you go and lie down until those impulses go away.

The reality is that there is no substitute for professional students taking apart and studying an actual human body. Like most people, I would prefer *not* to be treated by doctors who had never studied the human body for themselves, never put scalpel to human tissue before putting it to mine. I believe most thoughtful people would agree. The deficiencies of the alternatives to cadaver labs seem obvious to everyone except some politicians and some school administrators.

Of course, there are Anatomy courses in which the cadaver laboratory would not be appropriate. There are justifications for prosections, plastinations and virtual dissections in those courses, or as an adjunct to a cadaver course. Dissecting properly some parts of the body simply takes too much time for students to do themselves. It's impractical to require dissecting all the branches of the facial nerve on the cadaver's face (*Netter 5th ed.* 24, entitled "Facial Nerve Branches and Parotid Gland"), but a prosection can be instructive. My best estimate is that it takes three or four hours to do an acceptable dissection of all five branches, loosely categorized and recalled by the common mnemonic, "two zebras buttoned my coat." In most cases, your instructor will not expect you to do such an arduous task; they will

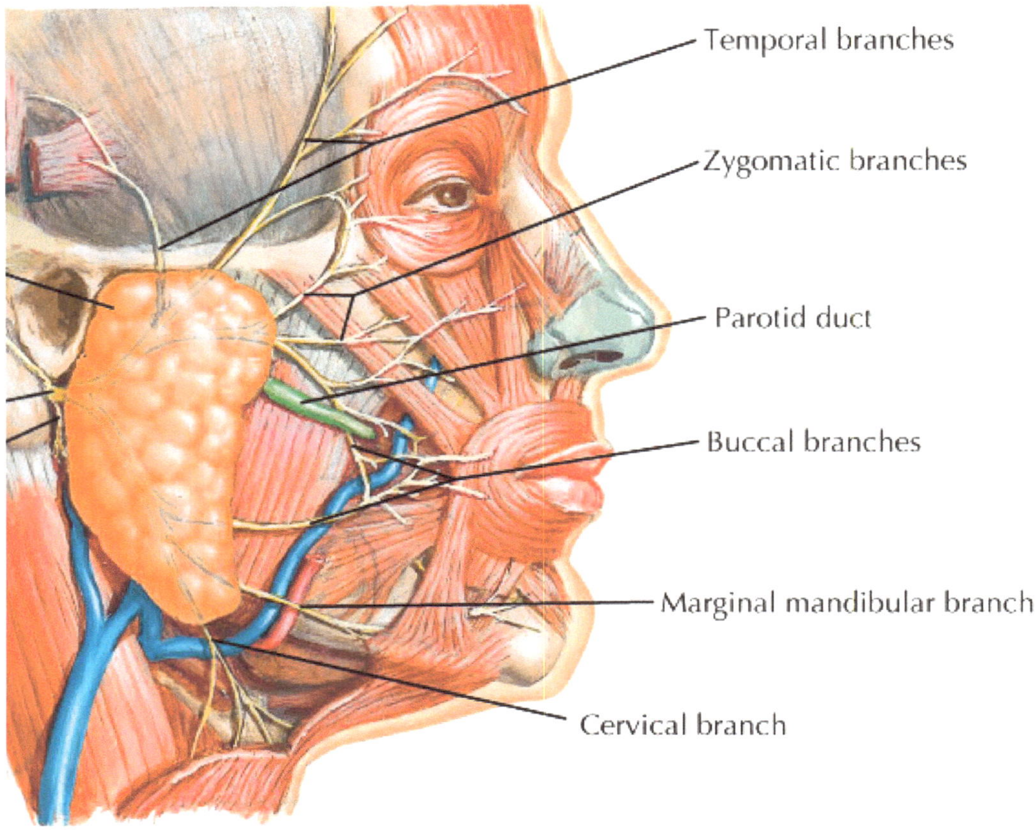

Netter Atlas of Human Anatomy, 5th Edition 24.

encourage you to find one or two representative branches of the nerve on the face, and, to get the complete picture, you can extrapolate your two nerves to a well-done prosection of the entire plexus. Alternatively, you can observe the plastic face models in the lab, or the drawings in your anatomy atlas.

An Anatomy Laboratory is expensive to run, so the reality is that university administrators, many of whom are naïve about Anatomy, will attempt to find less expensive methods to teach Anatomy. Those methods often vitiate the traditional laboratory experience. Typically, the exclusive use of these alternative methods fails, leading to the eventual resurrection of the cadaver lab. It seems to me that almost all such schools eventually go back to the cadaver lab. They're like the person who was delighted with the invention of a revolutionary new method of telephoning: the cell phone, but the person was constantly losing his. He struggled with the problem of misplacing it, over and over. Suddenly he had this amazing idea that it would solve his problem if his phone were attached to the wall, with a wire.

People say the study of Anatomy by ersatz methods is the way of the future, *and in my opinion, it should remain in the future: the distant future.* In the meantime, let's make our cadaver lab the best it can be.

The teacher named Dana in Bill Hayes' excellent book *The Anatomist* was discussing the best way to learn Anatomy. "...even every Saturday night. That's the way you learn Anatomy. You sit down with a dissection manual and a cadaver, and you just slowly go through everything."[4]

Let's take a look at some alternatives to the cadaver lab:

Plastic Models

Most academic anatomy labs possess a congeries of excellent plastic models, demonstrating various parts of the body. I highly recommend their use, to augment your cadaver laboratory experience. They have several distinct advantages:

1. They are several orders of magnitude larger than the natural structures.
2. They are colored brightly for easy identification: nerves in yellow, arteries red, and veins blue. Muscles are the color of uncooked steak.
3. They come apart, and can be explored from all angles.
4. They enable an easy familiarity with anatomical relationships, before going to the cadaver.

My recommendation is this: when you study any complex or miniscule structure, go to the plastic model before dissecting the cadaver. For examples, use models of the eye, the ear, the larynx, the infratemporal fossa, the spinal cord, the pelvis and the genitalia.

Be cognizant of the need to dissect the same side of the body as the plastic model represents. Otherwise, when you go to the cadaver, you'll need to construct in your mind the mirror image of the model. Or use your laptop to turn a photograph of the model to a mirror image.

There are other, similar attempts to simulate actual human body dissections. For one example, Syndaver models, which are models made of synthetic material. Syndaver mannequins and anatomical models are amazing, and have a place in the training of any healthcare provider. They are an excellent ancillary instrument, and are most useful in demonstrating medical procedures.

Use clean gloves to handle your models. Keep those models clean! Don't set them down on the donor body. Most of all, don't lose any of the parts of these plastic models. They are expensive, and if even a single piece of a model is misplaced, it can render the entire model pretty useless.

Prosections

It's just not practical for novices to dissect some parts of the body; the middle ear comes to mind. Nor is it practical to dissect the facial nerve plexus on the face. Nor the pterygopalatine fossa. In the case of these tedious dissections, most programs will do a few prosections: someone dissects the part, taking all the requisite time to do so, and then that dissection is made available to everyone in the class. This is a system that works well. However, the prosections won't last forever, and those areas must be re-dissected periodically. Some programs hire students to do prosections over the summer; others ask their instructors or teaching assistants to do them. Commonly, when you study the joints, instructors will trot out various joint prosections for you to study.

Prosections are also useful for review. You can go through them fast, like flash cards, and discover areas where you need to go back and study on your cadaver. It's wonderful to work through a prosection with a group of students. Think of it this way: I'm often called to a table with a well-done dissection by students. By the time I get there, it's a prosection, for me. Still, in my opinion, students will always learn the Anatomy best by doing their own dissections.

Plastinated Specimens

Not long ago a process called plastination was developed. Plastination removes water and fat under vacuum, and replaces them with polymers, which will theoretically last for decades without changing. This plastination process allows us to preserve prosections indefinitely. Of course, to the administration of a university, this looks pretty good: they will save money if they buy a plastinated specimen, because it will eliminate the need to acquire new cadavers each year. They might even consider using plastinations exclusively.

The disadvantages of exclusively using plastinations include:

1. Plastinations are expensive, and can cost as much as several thousand dollars for a single specimen of just one part of the body. There are therefore major security considerations for a library of plastinated specimens.
2. Even plastinated specimens don't last forever. They are almost completely inflexible. If you try to open a chamber of the heart, you may break the "door" off from the specimen. They can be repaired, but it is expensive and difficult. Parts end up missing, or broken, or crushed.
3. The laboratory sessions descend into a standing-around observation session, like judging pumpkins at the country fair, instead of a problem-solving, team-based activity. There is no doubt that plastinated specimens have a place in most Anatomy Labs. But they will never replace dissection for quality learning.

Virtual Dissections

Some schools are beginning to do Anatomy with virtual dissections: the students "dissect" by cutting and pasting and so forth on an image on their computer. I think the deficiencies in this approach are obvious. A computer can do the work of a dozen ordinary men and women, but a dozen computers can't do the work of one extraordinary man or woman (like you) in the Anatomy Lab.

Animal Dissection in the Human Anatomy Course

When a cadaver lab is unrealistic, one good option is a human anatomy course that dissects cats. There's a fine dissection manual written by Elaine Marieb and Lori Smith,[5] that guides the student through a cat dissection, with frequent reference to the human parallels. Some anatomy classes dissect a bovine eye or a porcine heart to augment or replace dissecting these tiny organs in a cat. These alternatives can be a real benefit in classes in which human dissection is not feasible.

Conclusions Regarding Dissection Substitutes

Many people unfamiliar with the cadaver lab (even school administrators) find themselves in the position of making important decisions about that lab. They sometimes fall prey to those ersatz types of labs because they perceive that they will save the school money, or diminish the time that students need to spend studying Anatomy. And they think the Anatomy Lab is scary, and messy. I like to say that cadaver dissections are like children: you don't really understand them until you get one of your own.

All these attempted substitutes for the cadaver laboratory bring to mind a quote from the novel *Pride and* Prejudice, describing a ballroom dance: "I should like balls infinitely better,' she replied, 'if they were carried on in a different manner; but there is something insufferably tedious in the usual process of such a meeting. It would surely be much more rational if conversation instead of dancing were made the order of the day.' 'Much more rational, my dear Caroline, I dare say,' replied her brother, 'but it would not be near so much like a ball."[6]

Making Anatomy Lab Enjoyable

Your instructors work hard at making the Anatomy Lab enjoyable. Efforts to invite friendly competition between tables are almost always helpful. If one table finds an excellent example of chorda tympani nerve, they are quite pleased to show it off to the rest of the class. Mock practical examinations, held at the end of some class periods, with each table setting up three questions on their cadaver, can be fun and can inform you of certain areas that need further study.

In my lab we give informal awards for a good job on difficult dissections. Generally, the awards are Snickers bars, bought by me in large quantities. They are *wrapped*, so they can be handed out freely in lab. For example, I awarded Snickers bars when a certain table dissected carefully the infratemporal fossa from the medial side (the perspective of the dentist), along with the traditional lateral approach. Another award was given, when dissecting the spinal cord, to a group of students who extended their dissection to trace the intercostal nerves to and through the dorsal root ganglion. An attempt at dissecting the pterygopalatine fossa might even earn you *two* Snickers bars.

We also gave Snickers bars for the largest gallstone, and another for the most beautiful gallstone, discovered in the Lab. The competition was keen in this endeavor. One year a student made a *sub rosa* trip to the parking lot, picked up a stone the size of your great toe, painted it green with a felt marker, and entered it into the competition. He was summarily busted, and did *not* take home a Snickers bar that day.

Another enjoyable aspect of the Anatomy Laboratory is when we are called upon to give tours to groups of young people. Most of them are high school students taking courses of study that can be enhanced by a demonstration in the lab. We are most happy to give them the tour. The vast majority display a dignified intellectual curiosity about the donor bodies. I always feel sorry for those who stand back and execrate the proceedings, because they are missing an opportunity to learn that doesn't come along every day.

You will find the Anatomy Laboratory to be comfortable, casual, and low-stress (for the most part). Over the course of your study, you'll find yourself feeling at home in a place you never thought you'd feel that way. Humor, and good humor, is welcomed in the lab, and is sometimes the glue that holds people together when things get stressful. The line between acceptable and unacceptable can be difficult to discern, and can vary from one laboratory to the another. If you're in doubt about the appropriateness of what you're about to say, don't say it. Keep in mind that any kind of japery and mockery regarding the donor body is completely out of place in any cadaver laboratory. It's offensive, like swearing in a kindergarten.

Halloween Costumes

Looking back on my long career as purveyor of Anatomy education, I believe I've had more humorous experiences than most. My! How I wish I could remember the name of the student, fully six feet tall, who presented himself in the Anatomy Laboratory wrapped from head to toe with purple yarn. He staggered, in the style of R2D2, to the front of the room.

"Who am I?" he asked, his voice muffled by the yarn.

I had no idea who he was pretending to be. The rest of the class was already in on the joke. "Pampiniform plexus!" he exclaimed.

Another time the class presented me with a Photoshopped picture of the pulchritudinous Meg Ryan with her arms around my neck, after I used a scene in *When Harry Met Sally* as a springboard for discussing why your nose runs when you cry.

I had the Best Body T-shirts

At my former school, at the end of each semester, we made an award to the table that did the best dissections overall, in the opinion of the cadre of instructors. This award was usually T-shirts, with the logo "I had the best body" in bold letters, and the explanation: "University of Detroit Mercy School of Dentistry Anatomy Laboratory" in smaller letters below. Every student desires a good, and good-looking, body, and I found that they worked assiduously for those shirts.

The Privilege

Try not to take for granted the great privilege of working in such an amazing place. Leonardo da Vinci said, "Experience is a truer guide than the words of others." I hope you will take this attitude to your experience in the cadaver laboratory. The inside of the body is a great mystery to most people. They entertain vague notions about where the organs are located, what they do and how they work. *But for those of us who have the splendid opportunity to explore it for ourselves, to actually see for ourselves the beauty and grace of the human body, and to learn its secrets in order to better serve our patients, it is an astonishing, compelling, once-in-a-lifetime opportunity*. I know that in your case, you won't take this unique opportunity for granted. I know you'll work hard and learn as much as you can. I know that's true, because you seem pretty wonderful to me.

CHAPTER 7A:
Discourse with the Donor

Very early in the morning, I was awakened by a slight brush of air across my face, accompanied by the soft sound of "whoosh," like that generated by a bat in flight. Then she spoke, and I opened my eyes.

"Look at me," she implored me. "Look at the meticulous workmanship in the dissection of my heart. Performed on a long Saturday morning in the laboratory by Dr. Midias."

"He's pretty adroit," I responded, feeling strangely rejected by her praise of my colleague, Midias. I thought, but did not say, that Midias could be insufferably sanctimonious. I tried a gentle criticism of his work. "He certainly has to work hard, to be good at dissecting."

"Habibi, he's not good; he's great. Several of your guys are good dissectors, but they can abandon any hope of competing with Midias," she said. She peeked again deep into her mediastinum, and said softly, "Dr. Midias is the augmented chord."

She rested her hand gently on the surface where her left breast used to be. It seemed to me she was feeling for a pulse, but I told myself that would be silly. I couldn't help noticing the clear distinction between the layers of intercostal muscles. My! That *is* good workmanship. I had to admit, the dissection of her chest was excellent. She seemed to read my mind when she said, "He simply puts the time in. You can't become a great dissector by rubbing a lamp. He stays into the crepuscular hours, and weekends in the lab are common for him. There's a reason the students call him 'Fastidious Midias' He simply works until he gets it right. He should have been a neurosurgeon."

I simply stared at her for a few seconds. "Maybe Midias should get a life," I said softly. I was hating this conversation. "I'm tired of his Excaliburish flourishes of the probe and scissors. It's just so *tedious*." I suspected that Midias was simply a reminder of my own wilting powers in dissection, dissecting having been *my* sweet spot for my entire professional life.

She could see that her remarks had offended me, and she tried to ameliorate them. "Well, O.K., maybe not a neurosurgeon. What I'm trying to say, Professor, is this: he works slowly, and he takes a break once in a while."

We thought about that for a moment, and I mumbled, "Some of us can't give it the time that we'd like. I've got six tables of students to instruct. On those rare occasions when I can take my time, I love dissecting like a potter loves the wheel."

She considered that for a moment. "Why is that? Why do some people like dissecting so? Is it somehow satisfying to make order from the chaos?"

"Dear," I responded too quickly, "we don't make order from chaos; the order is already there. Dissecting simply displays it. Shows it off, you might say." I loved teaching, even teaching her.

"You should pass that attitude on to the group of instructors who sit on stools in the corner of the lab. I swear that graduate student with the high-water pants and the white socks hopes that no one comes to call him over to a dissection table. He's not serious about teaching the students. When he does move about the lab, it's nothing but…*peregrination!*"

I nodded, and she said, "And Marcia is sweet on him."

"Marcia? How can you be so sure?

"When lab was over and she was leaving, he asked her where she was going."

"And?"

"She said she had to go study in the library. Then she said, 'You should come and study with me sometime.' He laughed. She laughed. There's something going on there. Graduate students and Teaching Assistants should not be flirting with students!"

She was clearly exercised about this topic, and she fairly shook at the thought. I could hear her denuded carpal bones clattering against one another. Then it was quiet in my bedroom for a long minute and I felt it necessary to bring some *sangfroid* to the conversation. I went back to the slacker professors.

"Those instructors have other things on their minds." I said, "They've got their research to worry about."

"Oh, now you're clearly prevaricating!" she shouted. "You don't mean that. Those instructors are just lazy. They sit on their stools over there, lined up like brooding chickens in a coop, and palaver endlessly. And when they do get around to dissecting, they make mistakes. Since they're unpracticed at dissecting, they avoid it, and they sit on their lab stools.

They're kids, yet they look like old men in their dotage, feeding bread crumbs and pocket lint to squirrels in the park."

I felt the need to come to the defense of my colleagues, (but not Midias). "Mistakes can happen to anyone. I myself, on my first day as a Teaching Assistant in the lab, mistook the ileum for an oversized vermiform appendix," I said. "It was a rookie mistake."

"Hah! Now that's quite a confession. How stupid did you feel?"

"I had stayed up half the night preparing for that laboratory session, and I had read that the appendix can be as much as ten inches long," I said. "I took advantage of the innocence of the students and I feigned great confidence. I found the cecum, and sure enough, there was a worm-looking structure extending from it. I naturally assumed it was the appendix. Turned out it was the ileum."

It was quiet in the bedroom for a moment, while we both pondered the predicament of a new instructor with a group of students looking on, mistaking the small intestine for a teeny rudimentary structure. She read my mind and said, "Now, I bet *that* was hard to explain away, with the students at that table."

"Yep. There was no covering it up. There was no smoothing it over. No way I could make a joke of it. At that moment, they all knew that their new instructor 'didn't know his gluteals from a fossa in the ground.' I laughed, and the students laughed with me."

"They weren't laughing *with* you," she said. "They were laughing *at you.*"

"I suppose you're right."

"Of course I am," she said. She could sense my embarrassment remembering my own inability to dissect such an obvious structure, so she rested the bones and muscles of her dissected hand on mine, and said, "Dude, it's called Occam's razor: the simplest explanation is probably the correct one. You applied that, and you thought it was the appendix. But you confessed your mistake, and kept learning. Those other guys have stopped learning Gross. They just haven't learned gross anatomy well, because they're anxious to get back to their research laboratories."

We agreed on that. I said, "Sometimes a person will achieve an advanced degree in Anatomy *without* learning Gross Anatomy well. Most research done for advanced degrees in Anatomy are concerned with the frontiers of learning: molecular anatomy, or cell anatomy, or pathology. Most of the research in Gross Lab has already been done, so it's not often that a graduate student can write a meaningful ground-breaking dissertation in Gross Anatomy.

That's why those guys never learned their way around a cadaver. And that's why they pass the time in the lab in persiflage."

"So *that's it*," she said softly. She pondered that, and said, "I suppose that will all be better when you stop dissecting cadavers and go to strictly prosections and plastinated specimens, won't it?"

I lurched to a sitting position in the bed. "What? Are you kidding? What's this about giving up cadaver dissections?"

"I'm concerned about prosections replacing dissection. I've heard the students talking at my table. Are you telling me you haven't heard those rumors yourself?"

"My Dear, if you don't ask me that question, I won't lie to you. But are the *students* talking about it?" I regarded that for a moment. "Are they in favor of it?"

"Some pro, mostly con. If prosections and plastinations replace dissection in the Anatomy Lab, people like me will pretty much disappear forever. Just the thought is traumatizing my muse," she said.

The thought occurred to me that *she* was traumatizing *my* muse. "No way. It's not happening. At other schools maybe, but not at this school." I put on my best unconcerned face. "Why don't you stop worrying, and let me get some sleep?"

"Really? I heard an administrator proposing just such a thing in the lab. She and her camarilla were talking about the expense of a cadaver laboratory, and the time it took from the rest of the curriculum 'in terms of each student's valuable time'." The wise cadaver paused. "Then she said that, in her opinion, their limited study time should not have to be spent dealing with dead bodies."

"That's the dumbest thing I've heard in a month," I said, fluffing my pillow, trying to appear insouciant. "This idea gets bruited about by school administrators every few years. Some professional schools are bowdlerizing their curriculums in this way, and the time that's devoted to cadaver study diminishes every year, it seems. But it's the students who suffer. Some Gross Anatomy faculty have never learned, say, the female anatomy adequately. Then they show up at a department meeting and they think the cardinal ligament is part of a small red bird. It's discouraging for us professors, because we work hard to keep standards up." I was becoming thoroughly depressed.

"You're criticizing the administration? You could get in trouble for that."

"I have nothing against administration, just against some of the people in it. Some administrators have no idea the things that are going on in their school."

She was listening intently, "The right hand doesn't know what the left hand is doing," she said. "And the big loser is the student. Is that it?"

She leaned over and looked at me sympathetically with her one undissected eye, and said, "Honey, we both know why they want to get rid of us cadavers. We also know that it has nothing to do with enhancing 'each student's valuable time.' It's not a pedagogical decision; it's an economic decision; it's about teaching Gross Anatomy on the cheap." She sighed. "I was a professor myself. I know these things. *Plus ça change.*"

She was correct, as usual, but before I could express agreement, she was gone.

CHAPTER 8: The Cadaver Laboratory Protocols

Procedures and Rules in the Anatomy Lab

In the cadaver laboratory, the most important principle of behavior is this: always be respectful of the cadavers. The opportunity to dissect a donor body is an immense privilege afforded only a few; it's the direct result of the donor's generosity. It was the donor's intent that you make full use of the opportunity to learn from their gift. Fulfil their expectations and honor their wishes by dissecting with purpose and gratitude.

After all, the reason you come to the cadaver laboratory is to help you better serve your patients. Our respect for the donor is so great that we refrain from all derogatory comments about the body. Neither do we make jokes about any part of the body. We don't display the bodies in any inappropriate manner. We don't use the laboratory computers for anything except laboratory business. Personal electronic devices such as phones and tablets are not permitted in the laboratory. Because there's no accounting for different individuals' tastes, no music nor ribaldry is permitted in the lab. And we do not discuss details of the cadaver experience in public, nor in any social media platform.

If you are indecisive about making a comment in the laboratory, use this criterion for evaluating appropriateness: speak of the donor body as you would if it were the body of a member of your family.

Dissecting a cadaver is usually the first experience a student has in manipulating human tissue. In that respect it is important: one can get the feel of bones and skin and other soft tissue. But bear in mind that dissecting a cadaver is very different from surgery on a living person. There is no liquid blood whatsoever involved in dissecting embalmed cadavers, although there may be some clotted blood in the heart and vessels. Virtually all the blood in the body has been driven out when a formaldehyde solution was perfused, under great pressure, throughout the body. And there are other differences: the contrast between consistency of fatty tissue in a cadaver and a living person is the contrast between hard jello and soft jello, respectively. Joints are stiff and sometimes immovable in a cadaver. Lymphatic tissue is scarce in the laboratory.

You will find that the process of embalming the cadaver under great pressure engorges the erectile tissues in the nasal cavity, and, yes, in the external genitalia. In the male cadavers, the priapism of the embalmed external genitalia can be shocking. Students are advised to keep their comments on a refined and adult level. Salacious or off-color humor has no place in an Anatomy Laboratory.

Stay the course, work with dignity and gravitas, even if, initially you find dissection of a human body difficult. As you proceed with your dissection, your donor body will appear less and less like a person, more and more like a fascinating educational specimen, and it will gift you with an enviable amount of information and understanding of the human body. Throughout your life, in thousands of encounters, you will fall back on anatomical information you learned while attending that donor body.

Attire

Most people wear scrubs in the lab because scrubs are cheap, and when the course is over you may want to discard them. Traditionally, the scrubs are green or blue, and the best ones fit loosely. Loose-fitting is good, because sometimes you'll encounter a fair amount of physical work in the cadaver lab. For example, turning the bodies can be ungainly, is always a group effort, and sometimes involves reaching across the body to the other side of the table. You'll want to be dressed appropriately. Some dissections require you to use saws and rongeurs. In my lab we provide yellow smocks of the hospital variety (open in back) that are worn over the scrubs; they may be made of cloth and laundered by the school, or they may be disposable (made of paper).

Sometimes a school or a hospital can be your anatomical haberdasher. I see students wearing scrubs bearing logos that indicate they're hand-me-downs from somebody's work in the operating room or the Emergency Department. Other clothes are acceptable, and many people wear blue jeans, T-shirts and sneakers. It's cool in the laboratory; you may prefer a sweatshirt, especially if it provides an opportunity to advertise your undergraduate institution. There's usually a locker room available; you may change clothes completely before and after the lab session. Most students don't.

Some dissectors expand their wardrobe by wearing shoe covers, plastic aprons, double gloves, surgical caps and hairnets. Some students have discovered the great benefits of wearing head lamps, like coal miners. Head coverings for religious purposes are acceptable. Masks with crimp-able thin metal bands over the nose are provided and can be worn, but this practice is usually discarded after the first session or two.

So you can see that it's not necessary to wear a suit of armor to dissect. Just wear something comfortable, something you're not going to wear anywhere else.

In my lab we rule against holey shoes. (That restriction is against *holey* shoes, not *holy* ones, whatever those are.) Shoes with holes and open toes, like Crocs, even though they're remarkably comfortable, can be hazardous in the anatomy lab. If you were to wear such shoes, there are times when cadaver juice can soil the top of your feet. Or think about accidentally dropping a scalpel on your foot.

Baseball caps are not acceptable, in my labs and in my lectures.

The Atlas

While dissecting, you should perpetually be referring to the atlas. *Don't waste time.* When you're not dissecting, take a minute and cover the labels in the relevant plate in your atlas, and quiz yourself. In my opinion, it's always a good idea to keep an atlas handy and open, while dissecting.

Instruments

Learn how to load a scalpel. It seems odd to write this, but here it is: when loading a scalpel, keep your fingers away from the blade. Pick up the blade with a hemostat and line up the diagonal back end of the blade with the diagonal ridge on the handle. Put the blade over the bump on the scalpel handle, and slide it on. And equally obvious: pay attention to what you're doing. My own experience is this: I was demonstrating loading a scalpel at a cadaver table with half a dozen students. I looked away *very briefly*, and, looking back at the dissection, I noticed that the cadaver was apparently bleeding. *Wait a minute! Cadavers don't bleed!* No, alas, it was my own blood, dripping from where I had cut my own left index finger. It was such a magnificent cut that I needed to go to the Emergency Room for eight stitches. I hadn't injured my hands that badly since I tried to open a wine bottle with a corkscrew and found out too late that it was a screw-top bottle. Do as I say, and not as I do: pay attention to what you're doing.

But my carelessness in loading the scalpel surely got the attention of my students. They were scrupulously careful in loading and unloading scalpels from then on. They even suggested that I should cut my finger every year, to demonstrate to each class the consequences of obliviousness.

Different schools will supply the student with different complements of stainless-steel instruments for use in dissection. Some schools require the students to buy their own kits of instruments for the lab. The ubiquitous probe has a screwdriver-looking end and a bent-pointer end. Various types of tissue pickup forceps are common: some tines are cross-serrated, and some have opposing interlocking "teeth." Several sizes and shapes of scissors are essential. Some instruments have spoon-shaped ends, for scraping, reflecting, and probing the tissues.

Usually some collection of the armamentarium described above will be provided by the school to each dissection table. Also available in the laboratory are reciprocating power saws, rongeurs (pruning clippers), machete-looking knives, and mallets.

Here's a tip for the Anatomy Lab, that I learned from Stanley Pride, president of Pride Manufacturing Company, a company that manufactured around 90% of the wooden golf tees made in the world. Stanley said when you come up to the green, leave your cart on the cart path and take just a couple of clubs with you to finish the hole. When you put one of those clubs down, you should gently put it in a conspicuous place, say, on the green. That way, when you leave the hole, you can glance back and you will see the club lying plainly on the smooth grass, and you won't forget it. If, on the other hand, you leave the club somewhere off the green, your glance back could very well overlook the club. You could arrive at the next hole, and say, "Hey! Somebody stole my nine iron!"

The same goes for sharp instruments in the Anatomy Lab: when you put down a scalpel, put it in the most conspicuous place you can find. Put it on the shelf under the window (on a white paper towel for contrast), or on a white paper towel on the cadaver table, or find someplace where the instrument is alone, conspicuous, and unlikely to be accidentally bumped. Be mindful of what you're doing when you put that scalpel down.

At the end of the session, dispose of the blade in the manner required in your particular laboratory. Some labs have a device that de-blades the scalpel for you: all you need to do is insert it into the device. Other labs require *you* to take the blade off (with a hemostat), and insert it into a "sharps container," usually attached to the wall. It's important to dispose of scalpel blades carefully. Several times, in closing up a cadaver that someone else has dissected, I have cut myself on a scalpel that was casually left somewhere around the cadaver. There is no excuse for leaving your scalpel in a dangerous place, and if you do, I assure you that your instructor will have no sense of humor about it.

The Diener

In every cadaver lab, there is at least one staff person who is responsible for maintenance of the lab. We call that person the diener (pronounced **dee** ner). *Stedman's Medical Dictionary* defines diener as "a laboratory worker who assists in cleaning; most commonly applied to laboratory workers who assist in the performance of autopsies and maintenance of morgues."[1] At a professional school, the term is commonly used regarding the caretaker of a cadaver lab. The diener maintains and inventories the instruments and specimens. He or she accepts and prepares the bodies when they first come to the lab, and performs similar functions prior to sending them back to the morgue. Typically, that person has performed all the activities in the cadaver course many times, and is an excellent anatomist and dissector. The diener can be of immeasurable value to the faculty, and to the students.

The word "diener" comes from a German word meaning "servant," but don't get the idea that these people are *your* servant. You'll be well advised to be *their* servant, by helping them in any way you can. Like the loan officer at your bank, the diener is in a position to make your life very good, or very bad. If you encounter that person transferring a cadaver from one table to another, offer to help. Sometimes turning a body over can be a three- or four-person job. A bit of help from you will engender good will between you and the diener. Remember, that person is in a position to teach you a great deal, and to make your life abundantly more pleasant.

The diener is a general factotum in the lab, including taking responsibility for cleanliness. Cleanliness around *your table*, however, is very much your responsibility, and each Anatomy course has penalties for being bohemian in this regard. In my lab, the floor looks like an ice hockey rink immediately post-Zamboni. Most Anatomy laboratories are the same. That's the way it should be, and your diener will insist on it.

Cadaver Table Dissection Teams

Unlike the folks in the Rembrandt von Rijn painting (page 153), cadaver table dissection teams are generally assigned randomly. It's important to get along with your tablemates, even if it's difficult. If you're assigned the same table as your truculent ex-fiancé, or the neighbor who intentionally poisoned your dog, or if the conversation during dissection is perpetually eristic, your instructor will probably understand, and assign you to a new table, especially if you ask early in the course. If, on the other hand, you want to change tables because you like another team better, or your tablemate insists on showing you photos of her vacation to Playa del Mar for a third time, or one of your tablemates is a recidivist passer of flatulence, your instructor will most likely tell you, gently, to engage your moxie, and get back to work. No matter what profession you're training for, occasionally you are going to be forced to work with some of the most incorrigible people on the planet. If you don't win the Cadaver Table Lottery, use that experience as preparation for life. Theodore Dalrymple, in his article *Letter to an Aspiring Doctor*, wrote the following:

> Whatever your inner state of turmoil when confronted by the immense showcase of human folly or unpleasantness, you must retain your outer equanimity, which does not come naturally and at first will take a mental toll on you. But habit will become character, and eventually you will learn to accept people as they are – even if they don't deserve it.[2]

Please don't misunderstand me: if your lab partners don't pull their weight, or if they say or do offensive things, or make advances toward you, or act in a way that could prove detrimental to your grade in the course, be sure to let your instructor know.

Here's what some slackers think about hard work at the cadaver table: "Say! I could gain an advantage over my classmates, grades-wise, if I didn't waste my valuable time dissecting. While my teammates are dissecting, I could walk around the lab and look at the other tables' dissections! Wouldn't it be smart of me to spend my time *learning*, while my table-mates are spending their time *dissecting?*" Those people, so over-conscious of their own excellencies, do not usually make good laboratory partners, nor good healthcare providers.

That self-absorbed person may think they're gaining a "leg up" on their peers by this kind of thing. People who crave to be the bride at every wedding, the baby at every baptism, and the dearly departed at every funeral should be made aware of the fact that they are not the first anatomy students to adopt this specious thinking. And they're not fooling anyone; what they're doing is as obvious as going the wrong way on the freeway. That person will alienate their tablemates, and, worse, they will alienate their professors. They'll become *persona non grata* throughout the lab. *And they won't learn the Anatomy nearly as well.*

Taking a cadaver course in Anatomy and eschewing dissection is like going to Italy and eating at McDonalds. There are serious advantages to actually performing the dissection, and there are serious consequences to shirking those responsibilities. Don't do it. Spend your lab sessions working together with your tablemates. There's plenty of time for sightseeing when the dissection is done.

Obviously, each member of a dissection team brings to the table (literally) his or her unique life experiences, talents and work ethic. Together, the team can achieve better outcomes, and learn more anatomy. While dissecting, you're surrounded by other students who have valuable knowledge. They will be able to explain concepts you don't understand, or suggest an approach that resonates with you.

Whether your particular posse is a study group or a dissection team, becoming one cohesive unit without adding or removing members, is more important than you may think. Pedagogically speaking, your group should stay with the same members, and work hard at cooperation. Bradley Staats, in his excellent book *Never Stop Learning*, extolls the virtues of working with a particular team, with the same members working together over a period of time. Even *adding* members can disrupt the cohesiveness. Staats describes how "throwing more person-hours at a software project slows things down considerably, but managers nevertheless see it as a viable solution when they fall behind. This point has been simplified as Brooks's Law, well known to anyone versed in project management: 'Adding manpower to a late software project makes it later.' We forget all the coordination that is required as we bring in new people."[3] Removing members of a group can have similar consequences.

Generations of smart anatomists have determined that the way to learn anatomy *best* is by dissecting cadavers, in small groups. Some schools have decided to increase, even double, the number of students at each cadaver table. I assure you, when that decision is made, it is not a pedagogical one, it is an economic one. It is made for the benefit of the school, and not for the educational benefit of the individual student.

Try to enjoy your tablemates, and do your best to pull your weight in the group's endeavors. This is what happens when one member of the group is irresponsible about doing their part: everyone in the group knows what's going on, and they seriously resent that person. At the same time, the slacker person begins to perceive the other members of the group as predators, and himself (or herself) as the predatee. The person begins to blame all kinds of difficulties *on the group*, which he or she perceives to be unfair. The person will quickly develop victim mentality. Do yourself a favor and don't be that person.

Care and Feeding of Your Cadaver

William Shakespeare wrote:

"When I have pluck'd the rose
I cannot give it vital growth again.
It must needs wither."[4]

Your cadaver also will wither, and it will do so quickly if you aren't punctilious in caring for it. Most everyone uses the same metaphor to describe the appearance of the cadaver: the musculature looks like beef. I suppose it does. If allowed to dry out, it looks like beef *jerky*. Anatomy labs use the cadavers for a variety of things: at my former school, the oral and maxillofacial surgeons study the heads, the orthopedists study the extremities, the Ob-Gyn aspirants search the female pelves, and so on. I call it "the economy of the Anatomy lab": the idea that one cadaver can serve a variety of people, each person with his own area of interest. In those cases, it's sometimes difficult to identify a specific person or group who takes responsibility for the body.

Cadavers come to us with the hands, feet and face wrapped in gauze, looking as if they'd tried to escape. The gauze actually protects the wrapped parts, it makes it easier to turn the body over, and it restrains the cadaver's limbs from being in the dissector's way.

In most Anatomy labs, there are no cadavers who suffered with HIV/AIDS, nor have they had hepatitis C. Even if they had, 10-20 gallons of embalming fluid have been pumped through them. Embalming fluid kills all pathogens. No viruses, no fungus and no bacteria can survive after the cadaver has been embalmed for two weeks. I have had cadavers in my lab that have been there as many as *seven years*. As long as I kept them moist with some sort of formaldehyde solution, nothing grew on them, or in them.

The good news is that no viruses, no fungus and no bacteria can live on a properly embalmed and maintained body. You needn't worry about getting an infection from your cadaver. It seems a little ironic that formaldehyde, the agent that renders the body so harmless, is a carcinogen. There's further discussion of this on the following pages.

You must not let your cadaver dry out. Therefore, at the end of each lab session, when you close the body and perform the housekeeping duties that I call "choreplay," find ways to keep your cadaver moist until you return to the lab. Your instructor will have definite ideas about how this is accomplished. In my laboratory, we use spray bottles to wet the body before covering it up. Each cadaver is endowed with its own beach towel, which is entirely moistened (not soaked) with embalming fluid. The beach towel is draped over the entire cadaver. We tuck the towel in at the edges, like tucking a child in bed. In most laboratories, this works very nicely.

You'll also want to keep your cadaver *clean.* If the contents of the large bowel spill onto another body part, clean it carefully. If possible, schlep it over to the large sink, and wash it there. If it's not possible to move it, get an abundance of paper towels, wet them thoroughly with water, and clean up the mess. And speaking of the mess, it's sterile, you know. It's not going to hurt you. Admittedly, that doesn't make it any less displeasing.

At the end of each session, you have work to do. Put *all your instruments* in the appropriate containers. You may want to clean them, or not, depending on the results of your dissection. Put all the parts of the donor body back in place, and fold the skin back over the work you've done. Replace the drape carefully over the entire body, and zip the bag. Doing so can be an arduous task – like putting together a piece of furniture from IKEA – but it is the right thing to do. This is especially important if you've been examining a cadaver not your own.

Studying Another Group's Cadaver

In most schools, students of Gross Anatomy have access to the lab 24/7, and I encourage you to spend some evenings or weekends there, with a group of students or by yourself. You can bring your dissection up to speed, and then even beautify it a little, and explore pretty much *ad libitum*, without the distractions of other students, nor the constraints of following a Laboratory Manual. Reviewing on a cadaver in this way will reinforce what you've learned, and pique your curiosity about things coming up.

In most anatomy laboratories, "usufruct" is practiced generously. That unfamiliar word is defined as "the right to enjoy the use and advantages of another's property short of the destruction or waste of its substance"[5] You have every right to enjoy the use and advantages of another group's cadaver, either during a lab session, or by yourself in the evening or on weekends. However, be aware that inductive reasoning has very little place in the Anatomy Lab. If we assume, as all good inductive reasoners do, that a tenth cadaver will be like the nine we've just dissected, we could be in for a big disappointment. They don't always follow the atlas plates, and they don't always follow each other.

We soon realize that there are anomalies, and differences between cadavers, and that realization gives us new appreciation for surgeons. Surgeons, also, don't always find the ideal anatomy atlas morphology in their patients. That's worth thinking about.

While it is entirely permissible to *inspect* the cadaver another group's table, it is common courtesy not to *dissect* another group's cadaver without the expressed permission of the other group. Most of all, it's important to put the cadaver right when you leave the laboratory: well moistened with embalming fluid, covered and zipped up according to protocols made clear by the professors in your specific anatomy course.

The Embalming Fluid

Most schools still use some derivative of formaldehyde to embalm their cadavers. In my experience, there are no alternatives that work as well. Formaldehyde is commercially available in a form called formalin, which contains 37% formaldehyde, 7% methyl alcohol (a stabilizer), and water. In my lab, we buy a 10% solution of formalin, and dilute that in half, and we use a spray bottle to wet the exposed surfaces of the cadaver. Then we use the same formaldehyde solution to wet some cloth material (a beach towel works nicely), and keep the cadaver draped with that cloth, inside a zipped-up body bag, between lab sessions.

Sometimes the embalming fluid will have some glycerin or other oleaginous substance, meant to make the solution smooth. That can present a hazard when embalming fluid is spilled on the floor: it's slippery. And the embalming fluid is an irritant in a cut. It is prudent to wash any cut with soap and water before applying a bandage or Band-Aid, to reduce this irritation.

Formaldehyde is a biocide, which is to say that it destroys living things. It's a disinfectant and tissue fixative. All anatomy laboratories in today's academic institutions are carefully monitored for preservative levels. Formaldehyde's aqueous solution is still the most commonly used preservative worldwide. You needn't be concerned about breathing formaldehyde in the amounts required in a typical Anatomy curriculum. Except for people who are compromised in some way (pregnancy, for example, discussed in the following pages), there is no need to be concerned.

Pregnant in the Anatomy Laboratory

The following is a review of the literature regarding the effect on pregnancy of formaldehyde: http://www.ncbi.nlm.nih.gov/pubmed/26375478.

The conclusions of the article include: "The reviewed literature contains data strongly suggesting that pregnancy can be affected by formaldehyde exposure. Therefore, on the basis our analysis, female students who might be pregnant should avoid formaldehyde exposure, including that in a gross anatomy laboratory. Instructors should find other means of ensuring anatomical competence for these students."

My personal opinion is as follows: In light of the fact that there is a great deal that we don't know about the relationships between reproductive health and the conditions in the Anatomy Laboratory, we must "err on the side of caution" in these matters. Most institutions are cautious about their liability, even if the student has a desire to attend. The student should be supplied with the literature on the subject, and then apprised of reasonable accommodations that the institution can live with. These accommodations may include delaying the student's taking the dissection part of the course, or offering alternatives to the long hours spent in the laboratory itself, or offering a respirator, or other options for the student. She should consult her healthcare provider and she may be required to provide written proof of those meetings. The institution will refer the student to the literature that is extant, and, after a careful reading, the student will be equipped to understand the school's decision. Some of those literature references are as follows:

The Centers for Disease Control has publications with recommendations for institutional policy:

http://www.cdc.gov/niosh/docs/99-104/pdfs/99-104.pdf
http://www.cdc.gov/niosh/topics/repro/formaldehyde.html
http://www.cdc.gov/niosh/topics/repro/jobHazardExamples.html

Reasonable people, even those who are not "with child," will agree that it's a good idea to avoid long exposure to formaldehyde. The institution can provide any student with information about the ventilation system and formaldehyde levels in the lab. Then the expectant student (and her doctor) can decide how they feel about it. Even if their decision is to attend the regular laboratory sessions, in my opinion it should be overruled by the school.

Mold and Yeast in the Anatomy Laboratory

It's disgusting to think that mold and yeast could infest a human anatomy laboratory, and, in fact those infestations are becoming uncommon almost to extinction, because we've learned to prevent them. Cadavers that are well cared for in an anatomy laboratory will resist yeast and mold.

Historically, a mold or yeast could spread through a laboratory with the celerity of roadside kudzu. I have witnessed such an attack, and its spread reminds me of a description of "amplification" that I read recently:

> If you take a bale of hay and tie it to the tail of a mule and then strike a match and set the bale of hay on fire, and if you then compare the energy expended shortly thereafter by the mule with the energy expended by yourself striking the match, you will understand the concept of amplification.[6]

Dr. Shockley's remarks reminded me of an event many years ago when a colleague phoned me to discuss his response to mold on just one of many cadavers in his Anatomy lab. The concept of amplification applies nicely to the potential for developing serious problems in his lab. Every effort should be made to "nip in the bud" any sign, even a suspicion, of yeast or mold.

Yeast and mold present differently on cadaver material. Yeast has a strong unique aroma, where mold does not. Yeast is generally colorless, while mold can be colorful. Yeasts are mostly harmless to living humans; molds are not. Either one is a big pain in the… Anatomy Lab.

If yeast or mold gains a foothold among the cadavers, it can be very resistant to efforts to eradicate it. A tiny bit of yeast or mold, left unchecked, can be amplified like Dr. Shockley's energy metaphor, and it can go through a lab quickly, with devastating effects. The keys to avoiding this calamity are basically twofold: the consistent liberal use of a formalin solution, and the proper application of same.

1. Cadavers must be sprayed with a formalin solution liberally at the end of each session. If it's a session of more than two hours, the cadavers should be sprayed at least once in the interim.
2. Between sessions, all cadaver material must be draped with a heavy cloth that is permeated with formaldehyde. We use *large* beach towels entirely wetted with

formalin solution. How wetted are they? If you were to wring them *real hard*, you'd be able to wring out a small amount of the wetting solution. That's how wetted.

3. The cadaver is enclosed in an impermeable body bag, which is either tucked in on the sides, or zipped up, to enclose the cadaver completely.

We have followed these protocols, and we simply don't have problems with yeast and mold. If you try to get tricky, or cut corners, or try to develop your own embalming fluid, serious trouble can follow. The embalming fluid smells bad, (so they tell me), but copious use of that particular fluid is effective in preventing yeast and mold.

The "Aroma" in the Anatomy Lab

When I started teaching Anatomy many, many years ago, the formaldehyde aroma was strong in the labs. When one entered the laboratory, the effluvia of formaldehyde was so strong that it seemed to be pushing the door open from the inside. Not so in today's Anatomy labs. With modern down-draft tables, air is drawn from the room at a high rate, and in most laboratories there is complete exchange of the air every minute or so. Furthermore, there are strict government regulations on monitoring the use of formaldehyde, and there is no longer any excuse for Anatomy laboratories to be as asphyxiative as they were even a few decades ago. In my youth we toiled ungloved and unmasked, in air so stultifying with formaldehyde that even hot post-laboratory showers could not remove the smell from our bodies. Those days are gone forever, thankfully.

Nonetheless, I periodically hear students complain bitterly about the mephitic vapors of formaldehyde in the laboratory. Some say it reminds them of low tide, or used ice hockey pads. Occasionally students will show up with half a roll of toilet paper stuffed up their noses. None of that goes anywhere with me. I like that scent. I must be a born anatomist, because the aroma of formaldehyde seems to agree with me. If I could, I'd buy a Scratch & Sniff book that expresses that aroma. Many anatomists share this view, so I'd encourage you to refrain from complaining to your professors about formaldehyde. Such intemperate gritching is offensive at worst, and bad form at best.

One reason that whining irritates the instructor so, is because the instructor has long since gotten over noticing the fragrance. Another reason is this: some of us believe that nobody has ever developed a better embalming agent than formaldehyde, so there is no sweet-smelling alternative.

My own mentor, a gentleman and talented anatomist named Edward Tracy, used to tell me that his family sometimes remarked about a certain aroma that followed Ed home. Now I wear that aroma as a badge of honor, in his memory, and in my own experience.

Be aware of a few things:

A well-cared-for embalmed cadaver is sterile, and smells a hundred times better than a cadaver that is not embalmed. I hope you get a chance to experience (once) the smell of a fresh (unembalmed) cadaver; they are out there, and in some disciplines they are necessary, but you will long for the formaldehyde when you finish your work with the fresh cadavers. Periodically, my colleagues at the morgue have done some experimenting, attempting to formulate an acceptable alternative to formalin embalming, but in my opinion, their work has a long way to go.

A bit of gentle advice: be discreet in your response to this. Some instructors are offended by drama regarding the aroma in the Anatomy lab. There is no sense in making a spectacle of elaborate measures to counteract the smell: "Where are the masks? Can't we get some Vicks in here to put inside our masks to cover up the smell?" These sorts of whingeing can be annoying to your instructors.

Dude: Get used to it. Take the same attitude as you did when your mother made you eat your spinach before leaving the dinner table: suck it up (literally and figuratively) and quit complaining. After a semester or two, and for the rest of your life, the smell of formalin will be a nostalgic experience, I promise.

I tell my students that I don't notice the smell, and I even tell them that I find the aroma of a nicely-maintained cadaver quite pleasant. "When my wife, who is also an anatomist, and I go on a date," I tell them, "I have her splash a little formaldehyde behind her ear. It's a real turn-on for me." I'm afraid that this simply confirms their conviction that "Dr. Forbes' mental faculties are slightly at odds with factual realities."

Barrier Techniques

Gloves are standard in the Anatomy Lab. For years, the gloves were made of latex, but many people have allergy to latex, and nowadays nitrile gloves are standard – the only gloves we have in my lab are nitrile gloves. And furthermore: you should wash your hands after each lab. Why? The same reason you wash your hands after surgery, even though, in both instances you've been wearing gloves. It's a good idea to wash your hands whenever you've been in any kind of operatory or laboratory, gloved or ungloved.

Students sometimes chasten me when I have a tear in my glove. I remind them that I learned Anatomy in the Jurassic Era, during which we would have considered the idea of wearing gloves like swimming in a snowmobile suit. In those prehistoric anatomy labs, we got along just fine. So don't worry if you have a tear in your glove and consequently expose your skin to the cadaver. Just replace the glove. The cadavers are sterile, they won't hurt you, and consider this: look how well I turned out.

In those rare occasions when mold or mildew is suspected, a person should change gloves every time they move from one cadaver to another. This requirement can become onerous pretty quickly, and unfortunately it is sometimes disregarded.

Masks are optional, and very few people wear them. They can be really annoying to the wearer, over a lab session of several hours. Some people put a dab of Vicks jelly inside their masks to overcome the smell of formaldehyde. I think the smell of Vicks is worse than the smell of formaldehyde.

I like to use loupes in certain phases of the laboratory dissection. Mine are the professional design, but there are less expensive options on the market, available on line, that can make your dissection of, say, the infratemporal fossa, easier on your eyes.

Smocks may be furnished by the school, or the school may require the students to buy their own. At my school, we have thin cotton smocks that we provide the students, and the school launders them. We encourage the students to wear the smocks several times between laundering. The important thing is to protect your "street clothes" from the formaldehyde smell. Instructors will typically wear a "white coat," although they're commonly not white. Pastel coats are fine, and they distinguish instructors from students. My school launders them for us, so members of the faculty are always well turned out.

Here's what I do: before I don the lab coat, I roll up the sleeves of my shirt underneath. The lab coats are long-sleeved, so the only formaldehyde residue will be on the coat, and not on the sleeves of my street clothes. You should do the same before putting on your smock.

Also, consider buying those nerdy pocket protectors for your laboratory attire, if it has a pocket. If you don't, the instruments you carry will soon wear a hole in the pocket and render it useless. I buy my water-resistant pocket protectors six at a time, because they inevitably become dirty. Some of the fancy ones even have space for a name tag.

Students are expected to wash their hands before putting on gloves, which is difficult, because wet hands, or even moist hands, can be troublesome to push into latex or vinyl gloves. We encourage the same ablutions again immediately after de-gloving. Most professional schools will make it a point to teach the hand washing techniques that they expect their students to practice, and some students practice those protocols in the Anatomy Laboratory. Brushes for scrubbing and picks for cleaning under fingernails are usually supplied. Most protocols will include the basics. The common steps are something like this:

- ☐ Wet your hands using warm, running water.
- ☐ Add an antibacterial liquid soap.
- ☐ Rub your hands vigorously for 20-30 seconds.
- ☐ Scrub with a brush all the surfaces of your hands and fingers.
- ☐ Use your fingernails to scrape the flat surface of your palm.
- ☐ Pay special attention to:
  - -the backs of your hands
  - -between your fingers
  - -under your fingernails.
- ☐ Dry your hands and use the paper towel to turn off the water.

Grading Dissections

I have found it unmanageable to grade my students on their dissections. There are simply too many variables: uniquely difficult cadavers, uniquely difficult tablemates, uniquely difficult situations. All of these result in the fact that it's hard to objectively grade one dissection relative to another. Objective criteria for bad dissections are difficult to quantify, but, like they say about obscenity, I know a bad dissection when I see one.

Neither can you say, "Good, better, or best" with any accuracy. Parameters for excellence in dissecting are vague, and must take into account advantages in environment as well as the finished product. Some cadavers are more amenable to dissection than others.

Ultimately, we depend on the individual student's performance on practical examinations for our summative evaluations. A school's most appropriate response, if you practice indolence or gross delinquency regarding the protocols of the laboratory, is to make your lack of professionalism part of your permanent record at the school. Such an assessment on your record is never a good thing.

Some schools threaten to take points from the grades of a cadaver group if their environment is messy or dirty. Attitude can influence your grade. And of course, points can be taken off for skipping laboratory sessions or for perpetual tardiness. On the one hand, I personally shudder at the thought of diminishing an *academic* grade on any basis other than *academic* performance. On the other hand, there is no greater motivation for students than the threat of reducing their grades. I therefore implore my classes to be conscientious in these points, and *puh-leeeeze* don't force me to adopt Mickey Mouse grading procedures in order to maintain professionalism in the laboratory.

Be Tidy

I don't think I need to remind you of this, but I'm going to do it anyway: be neat at your cadaver table. During my career as an anatomist, the condition of cadaver labs has evolved, to great benefit. When I began my long career in Anatomy, it was acceptable for labs to be smelly and shambolic. Instruments were sometimes left around willy-nilly, and in some instances the dissection tables looked as if someone had picked up the table and shaken it. Flotsam and jetsam cluttered the table and floor. There was a strong odor of formalin in the air. Spilled formalin and glycerin made the floors feel like you were walking on hair oil. There were dim overhead lights that flickered constantly, vibrated like a spin-dryer, and buzzed like hornets in a jar.

Those days are gone forever, thankfully. Nowadays, we take pride in our fastidiousness in the laboratory. The chemicals in the air and on the bodies are monitored carefully. Slovenliness on a cadaver table and its environs is unacceptable. Everything should be placed in the bins for which they are designated. I'm so fussy about placing instruments in the appropriate bins that my students call me "the bin man." (My wife calls me "the has-bin man.") The diener and the professors monitor the situation closely, and anything short of "clean and neat" can adversely affect you. This is as it should be; after all, the university is not your mother.

The Autopsy

I've had students ask me if dissecting a cadaver is like performing an autopsy. My answer is "not so much." One of differences is that a Medical Examiner weighs some of the organs. He or she does this with a scale similar to those hanging in the produce department of a grocery store. The weight of each organ is carefully recorded. You might be interested in those weights.

Bear in mind that there are 454 grams in a pound.

> heart: 320-350 grams (>400 is considered an enlarged heart.)
> liver: 1200-1500 grams

The Painting by Rembrandt von Rijn

This painting is on display in the Rembrandt Museum in Amsterdam. In 1632, when it was completed, the city of Amsterdam allowed only one cadaver dissection in the entire city each year, always in the cold of winter, owing to the unpleasant putrefaction of cadavers in hot weather. Since there was only one cadaver lecture allowed per year, the honor of demonstrating the dissections was given to an esteemed anatomist. In the case of the demonstration above, that honor was given Dr. Nicolaes Tulp, the City Anatomist. It's likely he had not done the dissection himself; probably he had employed students or apprentices to actually do the work.

Looking at this painting, the astute anatomist will observe the famous mistake that Rembrandt made when he painted this scene: the flexors of the forearm are shown originating on the *lateral* epicondyle of the humerus. Those muscles originate on the *medial* epicondyle. People who are more charitable than I am might give the artist the benefit of the doubt, and assume the demonstrator was pointing out an anomaly in the forearm of the body.

Also, there is a nerve-looking structure running on the medial side of the left hand, in an area where nerves are not commonly seen. Some have speculated that it's an anomalous structure: the accessory abductor digiti minimi muscle.

You noticed those discrepancies immediately, didn't you? Neither did I.

The demonstration was a public event, and probably there were wealthy businessmen as well as anatomists among the observers. This particular body was that of an executed

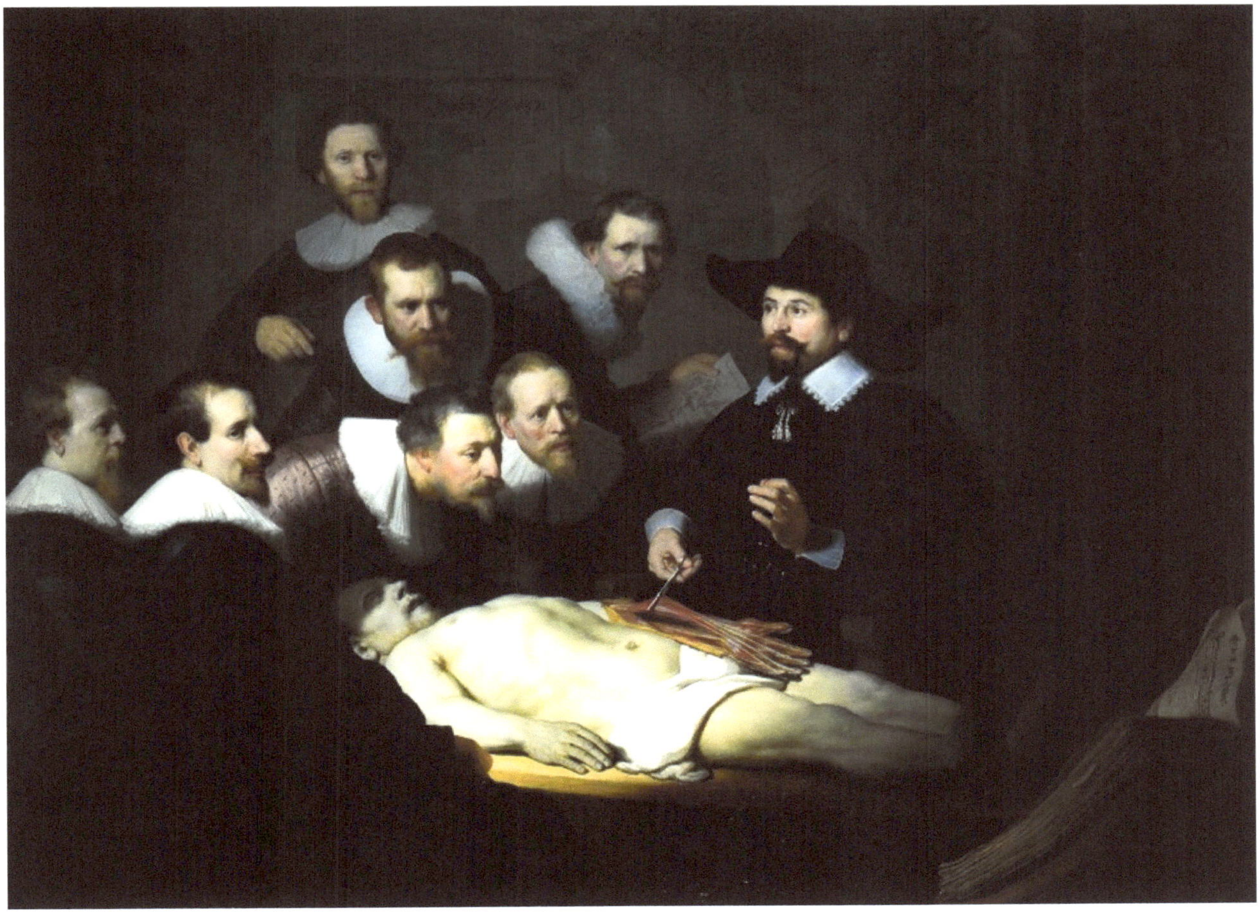

The Anatomy Lesson of Dr. Nicolaes Tulp, Rembrandt Von Rijn, 1632

criminal, Aris Kindt (alias Adriaan Adriaanszoon). The names of the men portrayed in the picture are listed on the piece of paper held by the attendance-taker at the back. The huge book at the lower right corner could possibly be Vesalius' book: *De humani corporis fabrica (The Fabric of the Human Body)*, written in 1543.

Notice in particular the gentleman leaning his entire upper body out over the body, thoughtlessly blocking the view of the others. He just may be the ancestor of a certain memorable student, who's been similarly inconsiderate in my laboratory.

CHAPTER 8A:
Discourse with the Donor

She was back. I glanced at the clock. "It's 2:30 A.M. in the morning!"

She smiled an indulgent smile. "'Two-thirty A.M. in the morning' is redundant," she said.

"Why do you do this? Why do you get up off the dissection table and intrude on my bedroom so rudely?"

She smiled a smile that was almost affectionate. All she said was, "Talitha Koum!" As was often the case in my conversations with her, I had no idea what she was talking about.

"What on earth do you mean: 'Talitha Koum?'"

"It's what Christ said when He raised that little girl from the dead in the book of Mark the fifth chapter, verse 41." She spoke with a distinctly offensive hint of impatience. "It means 'little girl, arise.' You should know this!"

"Oh, so I should know Aramaic, now?" But I was curious about her reference to the Bible. "Were you a Christian, then?"

"Not 'were.' I *am* a Christian, presently. Right now."

I blurted out my next thought before I considered it. "So you believe in the bodily resurrection of the believers? That makes no sense, coming from a cadaver. Look at you; you're in little pieces all over my lab. How can you be a Christian, and believe that we'll be resurrected *in our bodies?*"

She whirled around. "Of course I do!" she said. "Do you think bodily resurrection is tough for God? Did you ever read about Lazarus? He was four days rotting, I say *rotting*, in the grave, before he walked out. He was far more deteriorated than I am. If you think about it, putting me back together is not nearly the most miraculous thing about the construction of a person's body."

"Well, I don't know, I said. "How can you be so sure?"

"You think you know everything about the human body, and yet science can't even solve the basics. Like this: why is your yawn unsatisfying if you don't open your mouth wide?

Nobody knows. According to what the students tell me about your lectures, you must be an authority on observing yawning."

I understood her snarky response but did not acknowledge it. Instead, I tried to be agreeable. "You're right. I've never heard a convincing reason why we have to open wide for a satisfying yawn."

She went on, "We're talking about *God*, remember? The One who created the universe. Who created the electron. He created this beautiful body in the first place. Are you saying that… *repristinating* this body might be intellectually challenging for Him?"

I had a momentary thought that there were things going on in her mind that no living person could fathom. It was starting to give me goose bumps. "I'm not sure…"

"Habibi, a God that we can understand can't be God. You need to give this a little more thought. The reason that you find it incomprehensible is because you anthropomorphize God."

I felt like I was anthropomorphizing *her*. I said, "I have problems with some parts of the Bible, because I can't understand them."

"Ha!" she said. "When I was alive, I had problems with the Bible in the parts I *could understand.*" She smiled a happy, asymmetrical smile. She reflected a moment, and it was quiet in my bedroom. Then she said, triumphantly, "Sown perishable, raised imperishable…"

I said, "Do you really think God is going to put you all back together again, into a living form?"

"Of course. And even better than before. You and your scalpel-happy minions in the Anatomy Lab can't destroy this body; not really. Go ahead: try it; go crazy. You can't dissect me beyond the capabilities of the bodily resurrection that I'm *so looking forward to.*" She drew a deep breath, and shook her denuded boney finger at me. "You teach your students that the posterior cricoarytenoid muscle is necessary for life. As are the parathyroid glands. Necessary for life. O.K., go ahead and remove mine. I'll be just fine."

"I have no doubt that you will be," I said. She turned to leave, then whirled around and came back.

"I'm disgusted with people who come to the lab and are *so put off* by what you call 'the disarticulated bodies' like mine, that are simply lying on the tables. That's terribly offensive. Some of the thoughtless visitors you bring to the lab express their revulsion at the thought of exploring the body in this way, and that offends me. As far as I'm concerned, the lowest form of ignorance is rejecting something you know nothing about."

I thought about that. "I think that's brilliant," I said. "The lowest form of ignorance is rejecting something you know nothing about."

Her mood softened. "I don't mind so much a *visitor* being repulsed by the Lab, but the medical students? The first year of medical school is their a*nnus mirabilis:* they've been planning for this since they were twelve. Now that they're here, they act like it's gross."

"It *is* Gross," I pointed out. "That's what we call it: Gross Anatomy."

"That's what *you* call it, Cowboy," she said. "I call it home, at least for a few more weeks. In the meantime, you could have more respect."

"I respect you. I've built my entire career around you," I said, looking down. When I looked back up, there was only an empty, air-filled space; the wise apparition had vanished.

CHAPTER 9:
The Cadaver

Body Bequest, Organ Donation

Ours is a uniquely privileged study, and I'm perpetually reminded that our study is made possible only because of an unrequitable gift from people we've never met: the cadaver donors. They made a conscious decision to allow you to look inside their own deceased body. Those people had the foresight and the charity to give us the body that was their most intimate home for seventy, eighty, ninety years. For myself and my fellow Human Gross Anatomy travelers, the response to their generosity is profound gratitude.

Donating your body to science is a different matter from donating your *organs*. Organ donation is a huge network that provides miraculous remedies to suffering people. Body bequest is different: the person or the person's family make arrangements, sometimes years in advance, to "donate my body to science when I die." At their death, the entire body remains intact until it is embalmed and arrives in the Anatomy Laboratory. There, strict protocols are followed, and one body typically provides tremendous benefit to the education of a vast array of aspiring healthcare practitioners.

The Humanity of the Donor

When the cadavers from the morgue are delivered to the laboratory, they are not identifiable in any way. Their identities are withheld at the morgue and are linked to the donors by numbers, written and tagged appropriately, on the bodies themselves. A small amount of information accompanies each cadaver to the laboratory. That information includes the person's sex, age at the time of death, and cause of death (*cf.* page 169).

Cadavers look different from a living person: each body is uniformly gray and lacks the vivid colors of life that adorn the visage of a living person. Compared with a vibrant living human being, a cadaver wears a pallid monochrome.

The bodies arrive at the lab wrapped in muslin cloth, in pure white vinyl body bags with zippers. They are completely covered, and become dishabille only when the time comes for dissecting them. When we take those wraps from the cadaver, it can be unsettling to see the various evidences of humanity in, and on, the bodies. We often find nail polish on the fingers and toes. Sometimes it's bright pink, and lovely. The thing that unsettles me most is discovering a wedding ring, or a crease that formerly housed a wedding ring. Many cadavers have cicatrices or tattoos, but it always gives me pause when a donor body has a wedding ring crease. He or she has loved, and been loved in return. One can imagine the excitement the day the ring was first worn. If there's a crease for a wedding ring, we can be sure that the person wore that ring for many years. That particular reminder of the cadaver's humanity strikes me with emotions that I simply cannot describe. I personally know what it is to be totally devoted to a wife, and what that ring represents. Did this patient feel the same? Did he or she predecease their spouse? Did they have a family? Did he or she take care of this body in an effort to extend his own life, out of devotion *to them*? All accoutrements on a cadaver, and especially evidence of a wedding band, are solemn reminders of the *personhood* of the former occupant of that body.

We often see bruises on the bodies; ecchymosis and decubitus ulcers are common. The latter, commonly called bedsores, are usually found on the sacrum area, over the gluteal region, due to prolonged periods of lying in bed antemortem.

I've heard it said that there are 150 people in each Anatomy Lab session, *and that includes the cadavers.* That's incorrect; the reality is that the people are gone; what's left is the corporeal remains. We have the rare opportunity to examine those remains, and learn better how to treat our living patients. If you stop and think about it, you'll soon recognize the irony here: we're taking dead people's bodies apart, in order to learn to make living people whole again.

Tissues of the Body

Each body is made up of a combination of tissues: bones, muscles, nerves, organs, blood vessels and fascia*. But we soon learn the body is more splendid than merely the sum of its parts: anatomically, we're the same, and yet each of us is quite unique.

At the end of each laboratory session, we put the larger segments back together on the body. It can be a little like a life-size puzzle, the pieces fitting together a little less completely each time. Before we leave the lab, we save the small nondescript bits of tissue in a plastic bin (usually red), or a biohazard bag (also red). All human tissue is returned to the morgue at the termination of the course.

*Fascia: webs of densely woven tissue that enclose all the other structures. Looks and feels like diaphanous muslin.

Respect for the Cadavers

In some labs, it is customary to refer to the body as "the donor," or "the donor body," rather than "the cadaver." Some people consider one word slightly more respectful than another, and in our consideration of those bodies we want to be as respectful as we can. We want to recognize, in our references to them, the generosity of the person who donated his or her body to benefit us, and to benefit our future patients. We never use terms like "stiff" or other dysphemisms in referring to the cadavers. I myself am accustomed to the term "cadaver," and therefore I typically use that term in this book.

It's important to realize that the body that lies on the table before us recently housed a person who made the decision to donate his or her body to science. I cannot conceive of a motivation other than pure altruism, for a person to make that decision. The people who lived in those bodies were living people, and they may have held the door for you when you entered the grocery store, or maybe you saw them at your daughter's preschool. They were real people who made a real difficult decision as to their preferences for the arrangements relative to their own body after their death.

The cadavers in the Anatomy Lab invariably wear a serene expression. They actually *are* serene; the spirit of the person who occupied this body for many long years is gone, and what's left on the cadaver table is simply the human body that housed that generous spirit.

Occasionally, you'll be dissecting and suddenly you'll be startled by some detail that reminds you of that fact: *this cadaver once housed a living person*. Staples on the sternum and/or stitches on the heart itself will send your mind reeling into the nightmare scenarios that may have prompted those incisions. Tattoos are common, sometimes in surprising places. Implants, appendectomy scars, evidence of hysterectomy, all serve to reinforce that powerful fact that we always should bear in mind: this body once belonged to a person. That person cared about his or her body and likely went through massive efforts to keep his or her body alive and healthy, before they passed away. Therefore, we regard the bodies with the utmost respect.

There is other evidence of adversity in the life of your patient: a cicatrix, in particular, leads one to imagine what surgeries or accidents the person endured. Lungs can be especially dark and firm, which can be evidence of cigarette smoking (or longtime residence in certain cities). Stitches or staples inform you that you're not the first person to enter this body. Inguinal hernia repairs can be extensive and creative. Growths, bedsores, asymmetries, and other "barnacles from the sea of life" will inspire your imagination about conditions and events endured by the person who inhabited this body.

Care of the Cadaver

The group of students to whom the cadaver is assigned must take on the responsibility of caring for the cadaver in various ways. Primarily, we want to prevent it drying out. "Desiccation is death" to a cadaver. If any part of the body is allowed to dry out by excessive exposure to room air, it will become rebarbative and difficult to dissect, and it will be less beneficial to the student. The extremities are particularly susceptible to desiccation; therefore, in most labs the hands and feet are wrapped in some sort of cloth ("Turkish towels" or cheesecloth), giving the appearance that that they're being prevented from escaping. The hands and feet are kept perpetually moist with embalming fluid.

Students might consider covering the wrapped hands and feet with clear plastic (commonly a Zip-lock freezer container bag), to keep the moisture in. The head is kept wrapped in plastic, in a strong attempt to keep it moist until, and after, its dissection. And wrapping the face in this way makes the cadaver less person-like. After the dissection of a certain part is completed, it must continue to be kept moist, to enable review, and, later, to facilitate the formation of practical examination stations.

The novice dissector may notice some mildly alarming features of their cadaver. Hemorrhoids are common. Since hemorrhoids are basically *veins*, and since the embalming process puts fixing fluid into the blood vessels under great pressure, cadavers often reveal large hemorrhoids. You may notice that the external genitalia are erect. The same thing that makes parts of external genitalia erect in life (engorgement of fluid: blood) can make them erect in death (engorgement of fluid: embalming fluid). This is a good thing; erect organs are easier to dissect.

The classic *autopsy* incision at the Medical Examiner's Office (as seen on TV) is "Y-shaped." The two arms of the Y run from each shoulder joint, under the nipples, to meet at mid-chest. The stem of the Y runs down the midline, left of the umbilicus (to avoid the round ligament, which is on the patient's right), to the pubic region. In movies and television, the arms of the "Y" are often incorrect in this respect: in a female, they are commonly depicted running superior to the breasts. The correct autopsy incision should run under the breasts. And those arms of the "Y" in both sexes should extend to the acromion process, which is the highest point on the shoulder.

Your incision in the Anatomy Lab for removing the thoracic skin will be in a different configuration, as directed by your Dissection Manual. Your first incision will likely be horizontally along each clavicle from midline to acromion process. Your second incision is likely down the midline, extending along the entire length of the sternum, and then coursing laterally along the inferior border of the rib cage. It is carefully positioned immediately inferior to the rib cage in order to prevent cutting the diaphragm. It will leave the skin of the thorax with "two swinging doors" like those of a saloon in the old cowboy movies. Later on, you will make a similar opening on the abdomen.

In his "To be or not to be" soliloquy, Shakespeare spoke of death as "When we've shuffled off this mortal coil."[1] His point is that the trials of life (the "mortal coil") are over at the time of death. The cadavers in our laboratory are no longer concerned with the mortal coil. The generous donors who once occupied the bodies? Those folks are long departed. So let's work to overcome the initial trepidation we feel at removing the skin, and let's discover what's underneath.

It has been my experience that each donor body is often studied by several disciplines, and the labs work hard to use each cadaver to its maximum advantage. That is to say, orthopedists are interested in the extremities, Ob-Gyn residents focus on female organs, and cardiologists study the heart and blood vessels. These studies could conceivably occur all on the same body, if care and effort were expended toward that goal. It should be an encouragement to potential donors that their bodies have been used to such benefit.

And, if my college roommate becomes a donor, he should take satisfaction in the fact that his body, at least, finally made it into medical school.

Professionalism in the Anatomy Laboratory

To say that we treat the cadavers in a dignified manner is not to say that the Anatomy Lab is a solemn place. These lab sessions are quite casual, and a fair amount of good humor is generally enjoyed by people in the lab, especially after the first few sessions, as they get comfortable. (Of course, it gets a little tense prior to big practical examinations.) Here's a caution: practical jokes involving body parts are strictly to be eschewed. Remarks concerning the appearance of the body are kept on a clinical level. For example, one may be inclined to remark about the thickness of the adipose tissue (fat) on the back of the hips of a certain cadaver. But of course we don't denigrate the cadaver itself. Often and sincerely, we express our appreciation for the gift that this person made in donating their most precious possession so that we can further our knowledge. Every living person in the Lab understands that cadaver donation is a generous gift, not to be taken lightly.

Body Tags

The donor body will have an identification tag attached. The tags may be on the ear, or on a toe, or both. Of course, it is absolutely essential that you retain the tags on the cadaver. These will be used by the morgue to identify the body so the cremains can be returned to the family. If a tag loosens, comes off, or otherwise needs replacement, notify the laboratory manager or a faculty member.

Elderly and Obese Cadavers

A common complaint that students have about their cadaver is that the person was elderly at the time of death. I think those students are making a mistake. Consider this: if a person lives to be 96, they've had a pretty good body; wouldn't you agree? A superannuated body might make the *best* cadaver, actually. In cadavers that are "well stricken in years," you typically find all the organs in place, no major rearrangement due to heroic surgeries, and generally good anatomy over all. So I encourage students to appreciate the fact that their cadaver is "long in the tooth"; if the people who lived in those bodies were alive, they just might teach us all something about life, even as they teach us about death.

One of the advantages of a younger female cadaver is this: that body is more likely to have her reproductive organs intact. Hysterectomies are common nowadays, resulting in cadavers that are partially or completely lacking those organs. That can be a disappointment for the dissectors. Perhaps those students should get a refund on their tuition. (Good luck, on that.)

Another difficulty that students have with their cadaver is evidence that the cadaver had been a gourmand in life. Personally, I like the fact that an overweight or obese cadaver generally has big, strong musculature, and large structures throughout. They're like an NFL lineman: they may have plenty of adipose tissue, but along with it there are the big strong muscles that made this big body go. Sure, it's more work to remove the oleaginous adipose tissue. It's time-consuming to remove four inches of it from the hip region, for example. But when you finish, the structures underneath will be big, and strong, and easily identifiable.

Regardless of its weight, you will find that this cadaver you're dissecting holds an order and a beauty that you could not have imagined. It's a mistake to think about dissecting as "making order out of chaos." The order is already there, and if you're not amazed by it, you're not paying attention. Our job is to uncover that order; to clarify it, and to learn it. Cadavers manifest an order and beauty that is also within you, and inside your laboratory partners, and in the people you love, and even in the people you hate.

Why Donate?

What prompts a person to donate their body to science, to provide a cadaver at a professional school? One reason might be this: their action can, indirectly, improve the lives of many thousands of people. Think about this: a conservative estimate of the number of patients a doctor might see in a career of 35 years is 30,000. And a single cadaver educates five or six medical students. You do the math. It is a profoundly meaningful gift, with salubrious impact over several generations.

That's one reason that people donate their bodies to science. Others may donate because they or a loved one suffered with a disease of some type, and the person was persuaded to donate, in order to advance our knowledge in that area. Perhaps the person was a clinician, or their children were, and they honor that career choice with their donation. There are many reasons why people donate their bodies to science, and we are grateful that they do so. Some insights into why a person might donate his or her body to science is found in *Chapter 4a. Discourse with the Donor.*

Apparently there is no shortage of donor bodies. Generosity in this respect is common. Dental schools or medical schools that do not have their own morgues pay a fee for each body they borrow. They are transported by hearse, in a dignified manner, to the Anatomy Lab. At the lab, they are treated with high regard, and care is taken to keep the various components of a particular body together. At the end of the course, the whole body is returned to the morgue, where, in most cases, it is cremated. Then the cremains can be returned to the family.

Anatomy Lab is a defining moment for many aspiring healthcare practitioners. For much of their lives, their mental pictures anticipating professional school have involved the

Anatomy Lab, and the clinic bedside. Once they arrive, the cadavers never disappoint. They silently teach us, *even me*, despite the fact that I've enthusiastically been learning Anatomy since before you were born.

Turning the Body

In my lab, we try to avoid the common dysphemism "flipping the body," although that phrase is still used by some people. It refers to the action of turning a body from prone to supine or vice versa. We "turn" the body, thank you very much, and when we do, we make sure to get *plenty* of help.

When you need to turn a big, strong cadaver, get some big, strong living people to help you. Cadavers are the very manifestation of the term "dead weight," and if you drop one on the floor, it's a major job to get it back up onto the table. Believe it or not, it has happened. In addition to their weight, a cadaver is wet, and sometimes slippery, so it's difficult to pick up. And the floor can be slippery where the cadaver came to rest.

I've seen a number of instances where insufficient care was taken in turning a body, and the result can make you strangely uncomfortable, even shocked. There is nothing that will remind you that this body once housed a mortal being, more than the sight of a bungled effort to turn it over. So, when you turn a body, take some steps to ensure that you do it right. Check to make sure all the wheels on the dissection table are locked. Then get *more help than you think you'll need*, and figure out a plan before you begin. Tell everybody what that plan is. "We're going to pull the body toward me a little, and then we're going to lift my side, and roll the cadaver toward the other side. You, Joe, will mind the upper extremities, making sure they don't get trapped under the body. You, Rachel, will manipulate the lower extremities as the body turns. Ready? Here we go…"

Removing Cadaver Material from the Lab

In Michigan, where I worked, it is a felony to remove body parts from the Lab. Even though you might think it would be *really* funny, and an Achilles tendon might impress your athletic boyfriend, it's not a chance worth taking. Included in my Syllabus for all my courses is this portion of the Michigan Penal Code:

Please Read Carefully:
The Michigan Penal Code
Act 328 (Part)

AN ACT to revise, consolidate, codify and add to the statutes relating to crimes; to define crimes and prescribe the penalties therefore; to provide for the competency of evidence at the trial of persons accused of crime; to provide immunity from prosecution for certain witnesses appearing at such trials; and to repeal certain acts and parts of acts inconsistent with or contravening any of the provisions of this act.

History: 1931, Act 328, Eff. Sept 18, 1931
The People of the State of Michigan enact:
Chapter XXVI
Dead Human Bodies
750.160 Disinterment, mutilation, defacement, or carrying away of human body.

> Sec. 160. A person, not being lawfully authorized so to do, who shall willfully dig, disinter, remove, or convey away a human body, or the remains thereof, from the place where the body may be interred or deposited, or who shall knowingly aid in such disinterment, removal, or conveying away, or who shall mutilate, deface, remove, or ***carry away a portion of the dead body of a person***, whether in his charge for burial or otherwise, whenever the mutilation, defacement, removal or carrying away is not necessary in any proper operation in embalming the body or for purposes of postmortem examination, and every person accessory thereto, either before or after the fact, shall be ***guilty of a felony*** punishable by imprisonment for not more than 10 years, or by fine of not more than 5,000.00. This section shall not be construed to prohibit the digging up, disinterment, removal or carrying away for scientific purposes of the remains of prehistoric persons by representatives of established scientific institutions or societies, having the consent in writing of the owner of the land from which the remains may be disinterred, removed or carried away. ***(boldfacing added)***

You should be aware that dental restorations, such as gold crowns, are considered part of the body. Therefore, it is a felony to remove them from the Lab. My experience is that this Code is relaxed with regard to bones, as a rule. Bones, and especially skulls, have been transported outside the lab quite freely, without a problem that I'm aware of. In fact, at my dental school, we were blessed with a huge collection of natural skulls, and each student was encouraged to take one home to study for a semester. Obviously, those skulls were not harvested from our cadavers. Cleaning, sterilizing and preparing a fine natural skull or skeleton requires special care, and a great deal of work.

Be reminded that we are thoughtful in our care of cadavers *at all times.* When they are transported from the morgue where they've been embalmed, they are zipped into spotless body bags, and transported in a hearse. Two gentlemen in black suits respectfully deliver them on a rolling table through a private entrance to the lab. They are not transported in any area where the public is likely to be. The bodies are not identified by name; they are recognized entirely by a number on a toe tag and/or an ear tag. Their names are not available to anyone except the staff back at the morgue where they were embalmed.

All the body parts of a particular cadaver are kept with that cadaver throughout the course. The exception is tiny nondescript pieces of tissue. Parts that are removed from the body as part of the study are replaced within the cadaver at the end of the session. At the end of the Anatomy course, each cadaver is carefully re-wrapped, zipped into a spotless body bag, identified on the bag by the appropriate number of the toe tag and/or ear tag, and returned to the two gentlemen in black suits. They transport the body back to the morgue, where, in almost all cases, the body is cremated, and the cremains are returned to the family. If there is a reason the body cannot be returned to the family, the cremains are interred in a cemetery at a site designated for that purpose.

Cause, Mechanism and Manner of Death

There are differences between the terms "manner of death," the "cause of death," and the "mechanism of death." You will notice that knowledgeable people will be careful about which of those terms they use in describing a person's death.

> *Manner of death:* there are five manners of death: natural, accidental, suicidal, homicidal and undetermined. A deep laceration, for example, may have been accidental, suicidal or homicidal. The only natural manner of death is disease.

Cause of death: this is the trauma or disease that directly caused the death of the person. Drug overdose, stabbing, pancreatic cancer and heart attack examples.

Mechanism of death: this has to do with the body's response to the cause of death, that actually led to the cessation of the person's life. The cause of death may be heart attack, but the mechanism of death may be more specific: cardiac arrhythmia, for example.

The *manner* of death of the bodies brought to the Anatomy Laboratory is always natural death. The information we receive accompanying each donor body simply informs us of the *cause* of death.

Mortis

Rigor mortis refers to stiffening of muscles after death due to lack of oxygen. *Stedman's* puts it this way: "stiffening of the body, 1-7 hours after death, hardening of the muscular tissues... it disappears after 1-6 days when decomposition begins. Synonyms: cadaveric rigidity, postmortem rigidity."[2] Typically rigor mortis begins at the lower jaw or neck and spreads downward. It begins to show after two hours, and peaks at 12 hours, whereupon the whole body is stiff. In the next few days, the rigor mortis will be gone. For simplicity we say, "begins in six hours, diminishes in six days."

Algor mortis: lowering of body temperature after death (about two degrees Fahrenheit per hour at room temperature).

Livor mortis: blood pooling in the body due to gravity. It lends the tissue a maroon color. *Stedman's* defines "livor" this way: "the livid discoloration of the skin on the dependent parts of a corpse." On the same page, "livid" is defined as "Having a black and blue or a leaden or ashy gray color, as in discoloration from a contusion, congestion or cyanosis."[3] It's the "black" in a "black eye."

Now I hear you asking "What are 'dependent parts' of a corpse?" Dependent is a term that doctors use to describe gravity-related swelling in the lower body. Gravity has the effect of pulling fluid down toward the earth, causing it to pool in the lowest parts of your body, such as your feet, legs, or hands. If you're lying supine, your dependent parts would be those toward the floor.

Whoops! Touching the Cadaver

While I'm working in the Anatomy Laboratory, students sometimes point out *en passant*, that my glove is torn. They are concerned that I may get formaldehyde, or cadaver juice, on my skin, and I appreciate their calling my attention to it. I like to remind them that when I first studied Anatomy in the 1960s, only sissies wore gloves. We dissected with bare hands, never dreaming we'd someday wear gloves to dissect. Then overnight, latex gloves came on the scene.

At that time, dissecting with gloves on our hands felt like building a ship in a bottle while wearing hockey gloves. I've heard it said that "a man convinced against his will, is of the same opinion still," and that certainly applied to my group of colleagues. We preferred using our bare hands, because we were persuaded that we had better tactile sense that way. Eventually, of course, the pro-glove movement prevailed, and that's a good thing. Despite our objections, we were, and are, required to wear gloves when dissecting.

The gloves weren't nearly the same quality as the gloves we wear presently. For those of us who worked in those clunky gloves during the transition, "donning the cellophane" took a great deal of getting used to. Thankfully, today's gloves are of much better quality, and we've come to realize that no matter what kind of lab you work in, it's always a good idea to wear them.

If you find a tear in your glove, rest assured that a little cadaver juice and formaldehyde on your skin won't hurt you. The formaldehyde is a wonderful embalming agent, and kills microorganisms on a large scale. So if you touch a cadaver, or a "dirty" atlas, or some slimy instruments, simply wash your hands and don't fret about it.

In today's labs, I think some people may have gone overboard in the other direction, turning the air over in the lab every few minutes, and enforcing barrier techniques that remind me of the guy behind the bullet-proof glass, cashing checks in a tough neighborhood in Detroit. The bottom line is this: You should definitely wear gloves in the lab, but if you accidentally touch the cadaver *sans gloves*, it won't hurt you. It's simply not necessary to wear Viking gear in the anatomy lab.

Reviewing the Cadaver

I can't stress this enough: it's important to review repeatedly, in lab. Practice going through the structures that you see on the cadaver, and the structures listed on your Structures List, and those structures in boldface in your laboratory manual. Study like a musician does: practice, practice, practice. Space your practices apart. As much as possible, randomize the practices. Practice on cadavers other than your assigned one. *And be sure to review from both sides of the cadaver.*

Inductive reasoning will avail you in the Anatomy Laboratory. If you dissect a large number of cadavers and assume, as all good inductive reasoners do, that the next donor body will be like the ones you've just dissected, you can save yourself a great deal of work. However, it's not like changing a tire on your car. Cadavers don't all match each other *perfectly*, and they certainly don't all match the atlas plates. Each cadaver is the same as the others, and at the same time, each one is different. Your job, in preparing for your healthcare practice, is to learn the general pattern, and then to allow for minor exceptions to that pattern.

Making a good dissection is not a matter of rubbing a lamp; you must do the hard work of making it happen. If you're going to *excel* in the Anatomy Laboratory, you're going to have to go back to the lab to review, outside regular class hours, alone or with just a few students. Most anatomy labs in professional schools are open basically 24/7. You may be required to go through some sort of "Mother-may-I?" routine to get access to the anatomy laboratory, but it's worth the effort. If you happen to go there by your lonesome, you'll be surprised how much work you'll get done.

If you're unclear about what you're doing, review other cadavers that have already been dissected. There's nothing wrong with getting a second opinion.

Before beginning your dissection, do the following:

Take a blue or yellow smock, and write your name on it, directly over your sternum. *Take the smock off* while you write on it; otherwise, ink from the marker might go through onto your clothes underneath.

You must wear eye protection at all times when you are in the Lab. If you do not have eye protection, the school might supply you a set of goggles to wear. Select from the box over there by the gloves shelf. Those are goggles that former students have left, intentionally or unintentionally, in the Laboratory. You can return the goggles at the end of the semester.

Make sure that your cadaver has a toe tag and two ear tags, and that they all have the same number written on them. *Never* remove those tags from the cadaver.

Part of treating your cadaver with respect is this: when you're not working with external genitalia, cover them up with a towel or something. There are, of course, lab sessions in which we *dissect* those organs, and obviously we expose them during those procedures. But whenever it's practical to do so, it's just good manners to cover the private parts.

The Gift of the Donor Body

The cadavers in the lab are not necessarily unclaimed bodies, as they were when I was in school. In the 1960s, most of our cadavers were bodies for whom no one took responsibility: "unclaimed bodies." Some of the bodies came from foreign countries. We often wondered who they were, and where they lived, and how they died. Nowadays, we don't dissect unknown, unclaimed bodies. The practice of doing so exposes the school to litigation.

Therefore, today's cadavers are, for the most part, bodies of beneficent people who donated their own bodies to science (or their families did). When you stop and think about it, it's a remarkably generous thing to do, to donate your body to science. And it's a blessing that there are those among us who leave us this unrequitable gift, to study in the Anatomy Lab.

People lived in these bodies, assiduously caring for them, their whole lives. And in the future, if you become a healthcare provider, you will participate in caring for your living patients – examining them, advising them, treating them. Accordingly, in our work in the Anatomy Laboratory, we live out a major principle of the healing professions, and it's this: the reason we study donor bodies is in order to better serve our patients. The reason we travail in the laboratory is to equip us to better serve our patients.

The donor bodies in the lab remind us of the fleeting nature of life, and being so reminded, it prompts us to be kind, and considerate of one another. No matter what our stage of life, our time is short, and that realization often prompts us to be more circumspect, more thoughtful, in our discourse.

I've had the benefit of working in Gross Anatomy since before you were born, and I've loved working with people like you, who aspire to do work that is so meaningful, so interesting, so… en*gross*ing (!) I've literally looked forward to each meeting of the class. Gross Anatomy is indeed the floor of many rooms, and those rooms are physiology, pathology, histology, biochemistry, and on and on. If you become a provider of any type of healthcare, you have a truly enviable future: you will have the unparalleled opportunity of helping other people to better health. And you will have learned the *foundations* of medicine and surgery *in the cadaver laboratory*.

We begin working with cadavers as strangers – strangers to the human body and strangers to each other – and we invariably finish as good friends to both. I'm so glad you haven't missed this rare opportunity.

I like to reflect on my life in the cadaver lab. It has been my indescribable privilege to work with extraordinary students and a fascinating subject. I think that later on, when my own body is a cadaver in some academic setting, a strange thing is going to happen, when a group of curious students open up my thoracic cage. I envision them, five bright young people around a cluttered cadaver table, setting about their routines, just as they have done throughout the course.

They will no doubt be impressed with the difficulty of unroofing the heart, and removing the fat that surrounds that organ. They will struggle (as we all did) with finding the coronary arteries, because those vessels are deeply encased in fat. In observing my particular heart, some sharp-eyed member of the group will notice that my right ventricle is enlarged, a condition associated with my chronic pulmonary hypertension. Then, as I envision them, the whole group will suddenly stop what they are doing, and together they will gasp, and lean in for a better look. There, on the surface of the ventricles, they will see a startling sight. Those students will see many of *your names* written there.

CHAPTER 9A: Discourse with the Donor

When I started teaching Gross Anatomy, I wondered what a cadaver would think about a cadaver lab if a cadaver could think. Perhaps I spent too much time pondering that fantasy, because nowadays Cadaver Donor Number 2436, who lies on Table 6 in my laboratory, occasionally comes and interrupts my sleep. She has told me she was an English professor in life, and she's not loath to tell me what she thinks.

The dream I was dreaming was a pleasant one, until suddenly it ended with a shriek, like the sound of the needle on a vinyl record screeching to a halt. "Yi-yi-yikes!" she shouted. "The students at my table were *really* upset about their Midterm Examination grades!"

"I was having a great dream," I said.

"You were twitching and leering as if it were a naughty one," she said, smiling. Then she became more serious. "Hey, Professor, do you want my advice? You need to be a little more flexible in the answers you accept, on your practical exams."

I was surprised at her attention to this. I had always considered my examinations to be fair, and above reproach. "It's a hard examination, and it should be. It requires the student to come up with the correct answer, on the correct side of the body, spelled correctly."

"I'm not talking about propriety in the answers. I'm talking about fairness."

"I'm surprised. Give me an example."

"Monica had a question marked wrong, when her answer, inferior vena cava, was actually correct. The pin was in the inferior vena cava as it comes through the caval hiatus. You wanted 'caval hiatus' but the pin was in the IVC!"

"The IVC, at that point, is located *in* the caval hiatus," I said, weakly. "They *already know* the inferior vena cava. There is no reason I would test them on *that*, on this test. She should have known I was after the hiatus with my question. I just don't think she knows the caval hiatus from an eggplant, so she named something she *does know*."

"Not fair," she retorted. "Furthermore, I heard two of the students at my table complaining about getting the 'tendinous intersection of rectus abdominis muscle' question wrong, and for different reasons. But both of their answers were very close, and should have been given partial credit, at least."

"What? How can that be?"

"Ellen wrote 'tendinous interruption of rectus abdominis,' and Rudy wrote 'rectus abdominis tendon.' Come on! They know the anatomy – do they not? – and they got the question basically right."

"'Their answers are *descriptions* of the tendinous intersection; they're not the *names* of the structures."

"Oh, and Ellen wrote "transverse colon" when the pin was in taenia coli. *Taenia coli is part of the transverse colon!* The pin was in taenia coli; your question was 'identify the pinned structure,' and she named the transverse colon. How could that be wrong?"

"It's *wrong*, that's all. We studied the gross structures like the transverse colon, on their last exam; they shouldn't look for it on this one. The taenia coli is a small muscle *that's part of the colon.* The pin was in taenia coli."

"On the 'psoas minor tendon' question, Ellen wrote, 'psoas minor ligament'; a ligament is very like a tendon. Why isn't 'psoas minor ligament' O.K.?"

"No. A ligament is between two structures. A tendon is part of a muscle. They should know the difference. And who was the student who wrote 'medical umbilical fold' when the answer was 'medial umbilical fold'?"

"James. Come on. You're going to mark it wrong for a single letter? They're *medical* students, for crying out loud. They're thinking medical. It's a Freudian slip, and you should allow it. Medial, medical, fiddle, faddle."

"Nope. Spelling is important. There are plenty of kids who missed "splanchnic nerve" because they spelled it incorrectly."

"Doctor, it's not supposed to be a spelling bee. You're supposed to be testing whether or not they know the *anatomy.*

"That's just it. I'm tired of students coming to me early in the course saying, 'Dr. Forbes, I just can't spell.' They simply leave it at that, as if I should just shrug my shoulders and say, 'Don't worry about it. Just write whatever you'd like, and I'll mark it correct.' Well, you'd better *learn* how to spell, at least the anatomical structures."

It happens that this is one of my pet peeves, so I went on. "Remember, my dear, that they'll not just be writing in their charts; they'll be writing letters to other specialists, and giving depositions in legal cases. They'll be writing *prescriptions*, for crying out loud. When they're writing for Melatonin, they don't want their patients to get Metamucil. They need to be able to spell the structures correctly."

"Well, I think you're too hard on them. How about a little grace? You once told the class, '*Grace is treating someone better than they deserve.*' Remember? Some students can no more spell 'splanchnic' than they can kiss their elbow. I believe it was Vince Lombardi Jr. who said, 'Gentlemen, we will chase perfection, and we will chase it relentlessly, knowing all the while we can never attain it. But along the way, we shall catch excellence.' You expect perfection when you should accept excellence."

"When they're removing my gallbladder," I said stoutly, "I want perfection."

"But no credit is given for a student who gets it almost right?"

"The name of the structure must be absolutely correct, and spelled correctly!"

She looked incredulous. "You love the students, but you're resolute on this. Habibi, you're an enigma."

I was longing to end this conversation. "Yes, and I plan to stay that way," I said. Then I laughed. "I suppose I shouldn't have marked it wrong when Melvin identified the 'uvula' as the 'vulva.' Do you think I should have marked that one correct?"

She grimaced. "But when Melvin identified the deep inguinal ring as the "internal inguinal ring,' a little compassion seems appropriate."

"Here's the worst of it: at our Examination Review Session, where they can review their answers with the answer key, Monica erased her wrong answer, 'tail of the pancreas' and wrote in the correct answer, 'uncinate process of the pancreas.' Then she brought the exam to me, claiming that we had made a mistake in grading the question. If I would change the mark on that question, it would move her from 'passing the exam' to 'honors on the exam.'"

"How did you know what she'd done?"

"We make copies of the answer sheets immediately after the examination. We produced our copy of the original, and she was forced to admit her attempt to cheat. She'll need to meet with the Academic Performance Committee. We'll see what happens."

"It's difficult to take a charitable view of that one." The former professor sighed, and turned to go. Then she stopped, and stated in a stentorian voice, "One shouldn't be any kind of healthcare provider if one leads a life uncomplicated by the burden of conscience." She smiled and nodded in agreement with her own wisdom.

Then I watched her visage slowly dissipate, like a cloud of warm breath on a cold day.

CHAPTER 10: How to Dissect

Dissection is teaching Anatomy to your hands.

My Strongest Recommendation to You

My most important advice to my classes in Gross Anatomy is disregarded by too many of my students. It is this: come to Lab prepared. You really shouldn't go near a cadaver unless you have firmly in your mind what you are going to accomplish in dissecting it. If you're not willing to devote the time to prepare adequately, your desire for excellence in Anatomy is simply velleity. Here's the key: when we dissect, we're not *exploring*. Exploring is travelling in or through an unfamiliar area in order to learn about it.[1] By the time you get to dissecting, the anatomy shouldn't be an unfamiliar area; rather, we're uncovering and confirming what we've already learned is there.

You will see this in a few of your tablemates: oblivious and nescient, they don't know the purpose nor the sequence of the dissection. Consequently, they search back and forth, like an indecisive squirrel in the headlights, spoiling superficial structures in their effort to uncover deeper ones.

Try not to be that person.

Come to Class On Time

The Cardiology lecturer at my medical school locks the door when the class begins. In Anatomy Laboratory, we don't lock the door, but we do make announcements and instruction about the dissection, which is the *first item on the agenda of each lab session*. If you come late and miss those announcements, you're likely to encounter some serious drama in the Anatomy Lab.

Being tardy is a habit. If you have this habit and you allow it to continue in your study of Gross Anatomy, it can hurt you.

Follow the Directions

Over time, as you learn to dissect, you may decide to eschew the strictures of the dissection manual, and blithely "do it my way." Perhaps you consider it *infra dig* to follow a dissection manual in the manner of following a cookbook. You may consider your idea to be revolutionary in the world of anatomy dissection, and therefore you may feel *compelled* to do it your way, for the benefit of science. Most professors of Gross Anatomy would differ with you on this point.

For example, there really is "a right way and a wrong way" to open the infratemporal fossa, and if you wander too far astray, you'll have sectioned the ramus too far inferiorly, and missed the mandibular foramen and its wonders. It's important to follow the manual like a cookbook, unless directed otherwise. You'll find that your professors support *stare decisis**, and defer strictly to convention when it comes to dissection protocols.

**Stare decisis*: ("stand by the decision") the Supreme Court's principle that precedents should stand unless they are unworkable and unreasonable.

Another example is as follows: when we study the nose, we must remove the middle concha in order to see what's in the middle meatus. If we remove that middle concha without first using it to lead us to the sphenopalatine foramen, our search for that all-important "Gateway to the Nose" will be unnecessarily difficult and time-consuming.

I hasten to add that in most laboratories if you really want to follow your own muse in dissecting, there are ways to find those opportunities. We professors do that all the time: recently I found it interesting to dissect the infratemporal fossa on a hemisected specimen *from the medial side*, but I wouldn't encourage my students to do so, until after they've learned the anatomy the traditional way, by dissecting from the lateral side.

I hear you responding, "there are several ways to access the infratemporal fossa. In fact, two different professors recommended two different approaches to doing so." I submit that each of those two different approaches have been thoroughly vetted in years of their experience, and along the way they just may have discovered that your groundbreaking approach doesn't work very well.

If you'd like to try that innovative approach that you genuinely think would clarify the dissection for you, most instructors will make allowances for you to do so, perhaps even allowing you access to a separate cadaver. However, in my classes, that permission is contingent on two things:

1. You must perform your new innovative dissection on your own time.
2. You may not perform your new innovative dissection on the cadaver assigned to your dissection team. That would be unfair to the other members of your team.

We all learn something every time we're in the lab, you know, and that applies even to the professors. Here's a recent example: I'd always struggled with finding the round ligament in the female inguinal canal. It's difficult because the round ligament looks just like the surrounding tissue, and "blends into the wallpaper" of the inguinal canal. It's frustrating to have a group of students surrounding the table, watching you sift through a mélange of connective tissue, eventually being forced to conclude that this poor lady evidently never possessed a round ligament.

This past week I mentioned my problem to my colleague She shrugged her shoulders and said, "Why don't you simply open the abdomen and find the round ligament there? You can trace it from there into the inguinal canal, and you'll have it."

Turns out my esteemed colleague was exactly right. There was that round ligament, easy to find intra-abdominally, easy to trace as it sneaks surreptitiously down the inguinal canal. I had learned something important that day, as I do most days, when I practice the art of dissection. "Practice makes perfect," they say. I've been practicing dissection since before you were born, and perfection for me is still a long way off.

Tracing Vessels and Nerves

My colleague's advice comes in handy in many areas. When tracing a vessel or nerve, the dissector may tug on the structure in one area of the body, and observe the resultant movement in another part of the body. This allows the dissector to trace the course of the tugged vessel or nerve.

Here's a good example: if you dissect the oral cavity, and struggle to find hypoglossal nerve entering the tongue (many people do), go back to the neck where you've dissected it previously, find that hypoglossal nerve *there*, where it's obvious, and tug on it. While you're doing so, observe the floor of the mouth, and there you'll see the movement of the overlying tissue. The nerve is just beneath the moving mucosa. *Aha!*

That's the same procedure that I followed in finding that elusive round ligament in the inguinal canal. Tug inside the abdomen, and watch carefully the inguinal canal. You'll see it in

there, waving its little hand and saying, “Here I am. What took you so long?”
N.B.: Arteries, veins and nerves will survive strong tugging, and they will remain intact. Thoracic duct is the exception. It is unwise to do this tugging procedure with thoracic duct. If you tug on it, it will break!

The Armamentarium

If I take a minute to watch a group of students working at a cadaver table, one thing I notice is that the poor dissectors are quick to pick up the scalpel, while good dissectors do so rarely. Superior dissections are made primarily with *blunt* instruments. Early on, the observant student notices that instructors use a scalpel only as a last resort. When they do use the scalpel to cut a structure, they hold it gently, like a writing instrument, and not like a power tool. Rather than using the *blade* to *cut* the tissue, the good dissectors are sometimes inclined to use the *handle* of a scalpel, or another blunt instrument, to *spread* the tissue apart. Expert dissectors will insert the points of a closed scissors into the tissue, and then spread the scissors apart, separating one layer of tissue from another. Despite the fact that the word “dissect” comes from the Latin word “sectus” which means “to cut asunder,” experienced dissectors are judicious in using the blade. There’s too much risk of cutting something they want to preserve. Experts seldom cut; they prefer to separate with blunt instruments, or they employ the bluntest of the blunt instruments: their fingers.

It might be good to remind ourselves at this point: the organization of the structures of the body is already there. It’s beautiful, it’s amazing, it’s profound, *already.* Our job is not to create something; our job is to uncover something, to clarify something, to clean something, until the organization, which is already there, is clear.

Furthermore, inexperienced dissectors should bear this in mind: most tissues are already naturally separated by fascial planes. In dissecting those areas, sharp dissection is not necessary. You can test this by yourself: when you initially remove the skin on your cadaver, once you get a corner of skin separated from the body, insert your fingers into the subcutaneous tissues, push firmly, and see how easily the skin separates from the underlying tissue. You can do the same to separate muscles one from another. And it’s the same for most tissues in the body. Separate the tissues; don’t cut them.

In some institutions, students are required to buy their own instruments, but in other cases the school provides them. In the former case, the school specifies the “kit” that each student purchases. Different schools supply the student with different complements of stainless steel instruments for use in dissection. The ubiquitous probe has a screwdriver-looking end and a bent-pointer end. Various types of tissue pickup forceps are common: some tines are cross-serrated, and some tines have opposing interlocking "teeth." Several sizes and shapes of scissors are essential. Scalpel handles come in various sizes and shapes. Some instruments have spoon-shaped ends, for scraping, reflecting, and probing the tissues. My favorite dissection instrument is a wax carver that I carried over from dental school: a spoon on one end and arrowhead on the other.

Usually some form of the armamentarium described above will be provided in abundance to each dissection table. Also available in the laboratory are reciprocating power saws, rongeurs (pruning clippers), chisels and mallets.

The Bones

Every Anatomy Laboratory has skulls, full skeletons, and various bones around the room. Don't miss the opportunity to review those bones when preparing to perform a dissection. They are the framework on which all the other structures are displayed, and those bones will lead you like a GPS around the various parts of the body. When you're regarding the bones, be sure to look for evidence of structures passing through, or passing by: a groove on a bone (like the radial nerve's groove on the humerus bone), a "hook" over the course of a vital nerve (like the hook of the hamate bone over the ulnar nerve), and a roughness on the bone where a muscle attaches (like the medial side of the angle of the mandible, for attachment of the medial pterygoid muscle).

Find Something You **Do** *Know*

The story is told about the drunk who lost his keys while riding on the subway train. He was observed searching for his keys on the street level, under a street lamp. When asked why he was looking *there* instead of on the subway train, he responded that the light was better there.

Forbes Method of Practical Anatomy is as follows: in the cadaver lab, if you dissect out a structure and you can't identify it, don't go all wobbly and panic. Find something that you *can identify*, and work from there.

Finding something you *know* is getting back into the subway train, where the lost keys are.

Removing the Skin From the Cadaver

The first time they take skin from a cadaver, students frequently ask these questions: "Am I too deep?" or "Am I too shallow?" "How can I know when my depth is correct?" Here's what I tell them: in skinning the cadaver, you should lift the edge of the skin. If you can see the "cotton candy," (which is actually the subcutaneous loose connective tissue), you're at the correct depth. You will find that the skin will separate from the underlying tissue quite readily at that depth, even without the use of sharp instruments. In most parts of the body, *your fingers* will separate those tissues perfectly.

Once you have lifted a corner of the skin you're setting out to remove, cut a "buttonhole" in the skin, put your index finger through the hole, and use the buttonhole to pull on the skin, putting tension on the tissue underneath, and exposing it for further separation. You can put a surprising amount of tension on the skin, and the more tension you put on it, the more you expose the structures underneath.

Students want to know whether or not to completely detach the skin from the body. As a general rule, as we reflect the skin, we'd like to leave at least some small portion of it attached. Why? So that, at the end of the lab session, we can more easily orient it when we fold it back over the cadaver, to keep the body moist until the next lab session. However, in some areas (the armpit comes to mind), it's troublesome to leave the skin attached. If it will enable you to be more thorough in accessing and dissecting the underlying tissues, feel free to completely detach the skin from the body.

Whether you leave it attached or not, don't take the skin off in little pieces; keep each section as large as is practical. Push the detached or reflected tissue aside, and at the end of the session simply apply it to the body just as it was when you started.

Striving for Perfection

"Carter and Gray [illustrator and author of the original *Gray's Anatomy*] shared a passion, verging on the macabre, for dissecting."[2]

Most students who are new to dissecting come to the work with the desire for their dissections to be perfect in every way. Many of them have resolved beforehand that they are going to take whatever time and effort it takes, in order to achieve resplendent specimens of every artery, every vein and nerve, every jot and every tittle. (Actually, tittles are not body parts. Neither are jots.) These excellent students are aware that this is a once-in-a-lifetime opportunity, and they assume that the way to take maximum advantage of this opportunity is to commit maximum assiduousness to every aspect of each dissection.

This is not always the case. Skin removal on the torso and extremities, for example, is not a meticulous process, unless it's during surgery. Removing the skin from a cadaver is a matter of, well, *taking the skin off*. This is not the time for displaying how to treat your "first patient" with needless fastidiousness.

It's like worrying about whether the expression "anal retentive" has a hyphen: *who cares?* In taking the skin off the thorax, there's very little to be gained by the novice anatomy student if he or she takes four hours to do the job. So let's find the cotton candy, divide it as much as you can with your fingers, and let's get the skin off.

When you come to dissect the parts of the body that *do* require maximum effort and a great deal of time, let me give you some more advice: you can be as meticulous as you want, you can be an industrious artisan for an entire afternoon, but if you are not prepared in advance with a strategy for dissecting, and a goal clearly in mind, you will break structures you don't want to break, and you will not be happy with your dissection. How do you prepare? A typical preparation takes fifteen minutes:

Quickly read the dissection in the dissection manual (five minutes).

Learn the drawings in the dissection manual. Pick one drawing in *Netter*, or in another atlas. Cover up the labels and review the relationships (ten minutes).

Then, if time allows, another fifteen minutes:

Watch the appropriate Acland video (ten minutes).

Take a dissection photo atlas like Rohen's book, and simply look at the pictures (five minutes).

Now you're prepared to perform a "*Netter-rific*" dissection.

Mistakes in the Lab

Don't worry about making mistakes in the Lab. If the Anatomy Laboratory were a drinking game, and if a dissector had to take a drink every time he or she broke a critical structure, we'd be drunk all the time in Anatomy Lab. We're not. Don't worry about breakage; it happens to everybody. That's why God put right and left sides on a cadaver: if you want to redeem your ugly dissection, go to the other side, and start over. Fay Vincent, who served as Commissioner of Major League Baseball, said:

> If I spend a few hours with the young, maybe I can share some lessons I learned at bitter cost. Maybe I can talk of my mistakes and awkward stumbles. Maybe I can assure the young that wisdom is the daughter of failure yet the mother of success. I remember the acrid tastes of failures, but I quickly forgot the whiff of a success. [3]

Your success in dissecting, and in uncovering the mysteries of the human body, will be developed on the back of hard work and perseverance. If you aspire to be a surgeon, you can't learn to tie a knot successfully with one hand until you've failed to do so repeatedly. So persevere in your dissections. Wisdom is the daughter of failure, yet the mother of success. If you fail on one side, go to the other side, and try again. Mark Twain once said that a cat, having sat upon a hot stove lid, will not sit upon a hot stove lid again. But Twain went on to say that the cat won't sit upon a *cold* stove lid either. Don't be like that cat, in doing your dissection. Try again on the other side of your cadaver. You'll do better the next time.

Watch Your Professors

If your dissections look like a trailer park after a tornado, you may find yourself searching for ways to improve them. The best thing you can do to improve your dissecting skills is to watch your professors dissect. Yogi Berra famously said, "You can observe a lot, just by watching."

When I go to a table and offer to dissect a troublesome structure, quite often I notice that some students are turning away from the table while I'm dissecting, or standing back and talking amongst themselves. They simply want me to do it for them. This is a mistake. I'm not the best dissector, but I've been dissecting bodies since before *you* even *had a body*, and I can show you the solution to your inability to dissect this structure. It's not only courteous to pay attention; it can actually be educational.

When your instructor comes to your table, he or she probably has a half-dozen other tables that they're also responsible for. So the instructor will not be inclined to dilly-dally in dissecting. Furthermore, they have done this dissection a hundred times before. So you will be surprised how quickly they can dissect the tissue. Very often when I'm dissecting, I hear a student mutter, "Wow." That's not because they're impressed with my intelligence or dexterity (or my good looks); it's because of my celerity in dissecting; the circumstances of my employment make it necessary. I have other tables to get to. That's not the case when students dissect, so take your time. When you come to dissect, yourself, (I don't mean *dissect yourself*) you should work deliberately but not hurriedly. Rome wasn't built… well, you know.

Now let's talk more about dissecting in areas *other than removing the skin:* there, you will need to be cautious, comprehensively prepared, *and meticulous*. I've noticed that when inexperienced dissectors try to dissect those areas quickly, they grab a scalpel and cut (first mistake), rather than separating. Then, they hurry (second mistake), making the area of dissection look like it has been stirred with a stick. Furthermore, they neglect their laboratory atlas (third mistake), preferring to press on, essentially blindfolded, into the abyss. *Don't be that person*. If you err, err on the side of prudence. Take your time, consult your atlas, and give it the effort that it deserves, even if it necessitates your coming to Lab outside of normal class periods.

And beware the student, blessed with anatomical dauntlessness, who picks up a scalpel and says, "Hey, guys, watch this!"

Sharp Instruments

I clearly remember an instructor of mine who gave me two words of advice that have stuck with me *lo these many years*. "Dissect bluntly," he said.

As I've emphasized before, I agree: most dissection should be accomplished with blunt dissection. However, when you do use sharp instruments, make them as sharp as you can afford. Change scalpel blades frequently. If you are going to use the scalpel, bear in mind the well-known aphorism in cadaver labs: "You have a greater chance of cutting yourself with a dull blade than a sharp one." As far as I know, this has never been proven with double-blind empirical data; however, your humble correspondent can attest to the fact that it is true. Abraham Lincoln said, "Give me six hours to chop down a tree, and I'll spend the first four sharpening my axe." If you think about it, the sharper the blade, the more control you have, and the less likely you are to cut yourself.

Anatomy professors take care of their hands because dissecting is a big part of what we do. Consider a Human Anatomy professor on an elevator. The door is closing, and the professor hears someone holler, "Hold the door!" It's common for a person in that situation to wave his or her hand between the closing doors, causing them to reverse, and open, allowing the late-comer to come aboard. Not the anatomist. He doesn't jeopardize his hands in this way; he bends at the waist and puts his head in.

Look: scalpel blades cost around a dime apiece. Are you really going to injure yourself to save ten cents? I hope you are not. But how dull is too dull? As a general rule, it's best to change your blade every lab session, more often when you're using the blade more than you usually would.

N.B.: This is so important that I'm going to repeat what I admonished in Chapter 8, page 138: when you've used a scalpel in dissection, and you intend to lay it down, take a tip from golfers. If a person playing golf is approaching the green, he or she may take a 9-iron and a putter. While putting, the person will lay the 9-iron on the grass. Does that golfer lay the 9-iron *on the green*, or *on the fairway*? While using his putter, a good golfer will gently lay the other club *in a conspicuous place*. Common practice is to lay it gently on the green. That way, when leaving the hole, the golfer can glance back at the green and be reminded to pick up the other club.

Do the same with your scalpel. When you finish using it, you'll want to lay it down somewhere where it won't be in your way. That's fine, but *put it in a conspicuous place*, and put it in the same place every time you stop using it. I like to place mine on a white paper towel, for contrast with the stainless steel scalpel. That way, I reduce the possibility of brushing against it and cutting myself. And I reduce the possibility of forgetting it, and leaving it in the shroud when I close the cadaver.

When you close your cadaver at the end of a session, be sure to remove the scalpel and other sharp instruments from the environs of the cadaver. Several times in my career I have opened a cadaver after class, and in the process cut myself, on one occasion requiring stitching up. A good scalpel is sharp as a razor, and it needs to be handled carefully, and stored thoughtfully.

If you're observant in the Anatomy Lab, you will soon conclude that the good dissectors, especially your instructors, are very judicious in the use of the scalpel. Among your peers, you may lose style points by using blunt instruments when "a scalpel can get it done so much faster," but I'll bet your dissection will win the blue ribbon. Novice dissectors pick up the scalpel *to cut* the tissues. Experienced dissectors use blunt instruments to *separate* the tissues.

When they do cut, your instructors will likely use tiny scissors, using only the very tip of the tines to cut. They just "nip, nip, nip" as they section. Whenever they can, they put tension on the tissue by reflecting it, before they cut, for maximum visibility.

So when you do cut tissue, consider using the scissors. They allow a much more controlled cut than a scalpel, because you can *see* what you're about to cut. Place the tines of the scissors where you want them, then examine the situation critically. When you're convinced you're in position to cut the target, the whole target, and nothing but the target, squeeze.

Surgeons don't always use scissors the way normal people do. Instead of putting thumb and index finger in the scissors, they often put thumb and ring finger in the scissors. This leaves the index finger and middle finger free to rest on the top of the scissors joint, resulting in greater precision in placing the tines. Try it; I think you'll like it.

One final note: you can use the scissors in the opposite action for which they were made, to *separate* the tissues. Stick the closed scissor tips into the tissue, and hold them there while you spread the tips apart, thereby spreading and separating the tissues.

Tying and Sectioning the Bowel

Let's say you're tying off the bowel in order to remove it. You'll want to tie it tightly with string *in two places an inch apart* at the duodenojejunal junction, and *in two places an inch apart* at the sigmoid-rectal junction. Then, in each instance, cut the bowel between the two places you tied off. This procedure prevents spilling the scatological contents of the bowel.

In placing the string, I see students struggling to push the string around the backside of the bowel, because the gloves are slippery and the string is floppy. Here's the easy way: push a needle forceps behind the bowel, and grab the string with it. Then withdraw the forceps, pulling the string through.

After you've tied off the bowel and before you cut the tissue, you'll want to cut off the long ends of the strings. However, I strongly encourage you to leave the strings long, until after you've cut the bowel, and use them for manipulating the tissue. After the tissue is separated, *then* cut off the long ends of the strings. Clean the small amount of coprology from each cut end with a thoroughly wet paper towel.

Tie the string in the following manner: loop one end of the string around the other, pull it tightly, then press on your half-hitch with the index finger of your right hand. Pull the knot tightly with your left hand, and maintain tension on the string. Now continue to press your right index finger firmly on the knot, while you loop the strings around one another again, in a second half-hitch, to complete the knot. It's always a good idea to loop it around a third time, just to be sure. And while you're at it, contemplate this legerdemain: some surgeons can tie a knot with one hand.

Believe

Let's say you're looking for obturator nerve on the posterior body wall. It lies very reliably between the psoas major muscle and the vertebral column (*Netter 5th ed.* 260, entitled "Nerves of the Posterior Abdominal Wall"). But it's in there deeper than you'd think. It's like dropping your pencil in the car, and realizing that it's gone down between the driver's seat and the center console: it's hard to see, and even harder to retrieve. It can be so elusive to see that you conclude it's simply disappeared. Searching for obturator nerve on the posterior body wall can be a similar experience: if you don't really have confidence in your dissection, if you don't really believe it's between psoas major and the vertebral column, you may give up on obturator nerve before you probe enough. You'll say to yourself and others, "I guess it's just not there."

What's the problem? You've got to *believe* that it's there. That confidence will motivate you to keep looking deeper, and wider. Keep looking between the muscle and the vertebrae until you find it. Take another look at your laboratory atlas, check out the obturator nerve at another dissection table, take a minute to gather your thoughts, get a drink of water or a cup of coffee. Then, when you go back to your problem area, spread the tissues farther apart. Build your confidence in the fact that the structure is where Dr. Netter says it is.

Another approach: find the obturator nerve in the pelvis, where it is obvious, *tug on it*, and trace it back into the abdomen.

If you don't believe the structure is there, most likely it won't be.

The Joys of the Atlas

A person who works with his hands is a craftsman. If you allow your brain to teach your hands to be an anatomist, you'll be an artist. Be an artist, in the Anatomy Lab. Love this. If you make the decision to love dissecting, you will become an *anatomy artist.* You will get great satisfaction from your achievement in the lab. And I remind you: your indispensable guide in this worthy endeavor is the anatomy atlas.

I'm amazed at how students come to the lab and attempt to dissect without the benefit of a nearby atlas. Novice dissectors should always have one of the very good atlases *open*, available for easy reference. I admit: that atlas is going to get "dirty" (translation: smell like formaldehyde), and therefore it will need to be "dedicated" (translation: for use *only* in the Anatomy Lab). That copy of the atlas is not the one that you will take home and study at the dining room table. However, you can purchase a good anatomy atlas cheaply: a 2nd edition of the *Netter Atlas of Human Anatomy* on Amazon today costs eight dollars and sixty-five cents. You can get a previously owned *Grant's Atlas, the 10th Edition*, (the atlas that I love, and use, and refer to often), for three bucks plus shipping.

So an older edition of a good human anatomy atlas is not a huge investment. Put it directly on top of your cadaver, and refer to it often. Don't worry if it gets completely wet with formaldehyde; you can buy another one next year. And you can buy another copy to keep, and put on the shelf in your clinical practice, and refer to it for the rest of your life.

Take a Break

Have you ever done a very difficult jigsaw puzzle, one that challenged you a great deal? Have you ever completed a 1000-piece jigsaw puzzle? No? Then I'd like to kindly ask you to leave those of us who have, alone for a few minutes. Just skip over the next two paragraphs and then go on with your reading.

There. Now that we jigsaw puzzlers are alone, let me ask you: have you ever searched for a particular piece of the puzzle with heroic resolution, and eventually given up? It happens to me, and I think a reasonable person in that circumstance concludes that the elusive puzzle piece is on the floor, or missing from the box, stolen by another participant, or it's been eaten by the dog. It simply isn't there.

It's a good idea to give it up for a while. Have a cup of coffee, or a nap. When you return to the puzzle, there is a good chance that you will find the piece you are missing. It was there all the time, but you were approaching it from a different viewpoint, or under different assumptions, or you just needed a fresh look. The same is true when looking for structures in the human body.

Let's take chorda tympani nerve for an example. Remember that chorda tympani joins lingual nerve within the confines of the infratemporal fossa. You may carefully search up and

Autonomic Nerves in Head

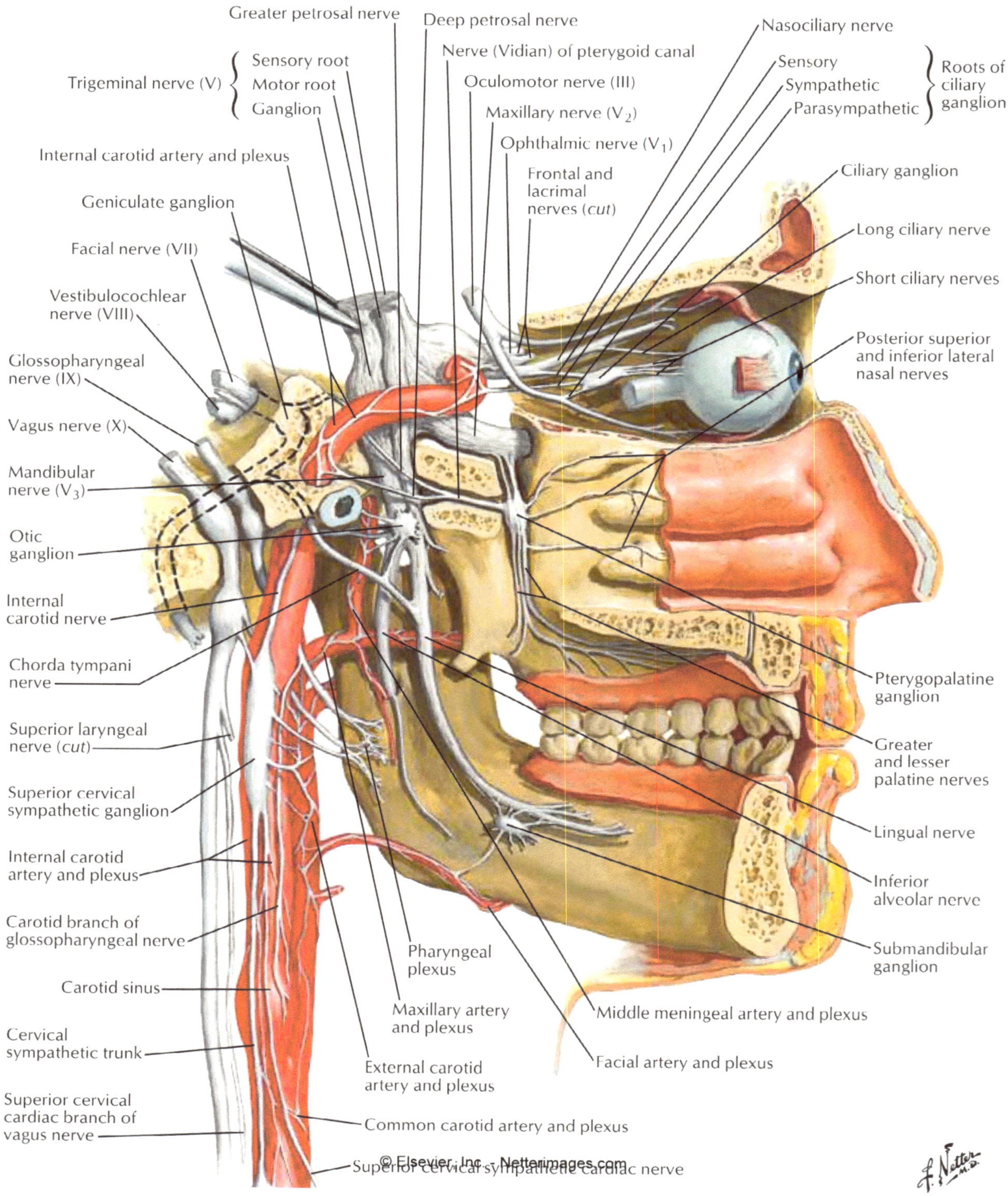

Netter Atlas of Human Anatomy, 5th Edition 130.

down the lingual nerve without finding it (*Netter 5th ed.* 130, "Autonomic Nerves in Head"). Eventually any reasonable person will conclude that chorda tympani does not exist on this donor body. You begin to think that when this person was alive, they obviously had no sense of taste on that side of the tongue, nor was there sufficient saliva being produced below the plane of occlusion on that side. The chorda tympani nerve simply isn't there. Here's what to do: wrap up the cadaver, get some lunch, and return. You'll be surprised what you'll find.

When I was a new instructor, I had this experience, and when I returned from lunch I realized I had been assiduously searching along the *inferior alveolar nerve*, not the lingual nerve. It's no wonder I couldn't find chorda tympani. (Please don't tell anybody about this.)

Or you can delegate the task to the entire small group around your table. That's what lecturers do, you know. Even the fancy lecturers, the experts, when they run out of meaningful things to say, they break the audience up into small groups for some Mickey-Mouse exercise. They get members of the audience to talk to each other. If you're frustrated with your dissection, engage the whole dissection table in your search. If two heads are better than one, imagine what five heads can accomplish.

Incidentally, if you're studying to be a dentist, please know that the same phenomenon can apply to removing a particularly difficult tooth, or locating the elusive fourth canal in performing endodontics on a molar tooth. If you work at it awhile and don't get the results you want, consider getting up, leaving the room, perhaps drinking a cup of coffee, before your return to your work. While you are gone, two things can happen. First, you will reflect on the project and return to it refreshed. Secondly, in the case of removing the tooth, there will have been some bleeding in the periodontal space resulting from your previous efforts, and inflammation will have developed, which tends to loosen the tooth, and you may find it easier to remove.

In the cadaver lab, whenever you are frustrated, take a break and come back. You will find that even your aging professors follow this approach.

"Appreciate" and "Reflect"

These two words are used in an unusual way in Anatomy. To "appreciate" structures in the human body is to study them. A great deal of what we do in the Anatomy Lab follows this pattern: we make elegant dissections one day. Then, the next session, we find it necessary to destroy that dissection in order to access and dissect deeper structures. However, here is the key: we must "appreciate" (translation: study) the initial dissection before destroying it. Once we've destroyed it, it's essentially gone, so we must squeeze out all the goodness we can, before moving on. And when we "reflect" tissue in Anatomy, we separate it from the underlying tissue. Don't make it any more complicated than that.

The Fleeting Nature of Your Beautiful Dissection

Let's imagine you've made a precise, meticulous dissection of the parotid gland, with all the tiny tendrils of facial nerve emerging anteriorly over the mighty masseter. You prepared well for this dissection, then you spent the time you needed, and it turned out first-rate. Here's the regrettable aspect of your beautiful dissection: its life is short.

The very next laboratory period, you'll be required to section some of it, in order to dissect the structures located deeper: the ramus of the mandible and the infratemporal fossa. It's an inevitable aspect of dissection, but a sad one. Your masterpiece dissection becomes as worthless as your masterpiece painting is since it fell off the wall.

Get Comfortable in the Lab

In her excellent book, *Body of Work*,[4] Christine Montross describes her experience after a month in Anatomy Lab. She writes, "We are increasingly able to detach from the strangeness of our actions, which serves us well."

A lay person taking a wrong turn and mistakenly entering the Anatomy Lab will immediately relate to "the strangeness" of the actions of the students dissecting cadavers. Those students that the stranger sees working have succeeded in "detaching" from that strangeness, and it is serving them well. For some, it has come easily. For others, it takes time and determination. But it is always worth the effort.

Cadaver dissections are like children: many people don't like them until they make one of their own.

Experts

When you finish your Anatomy course, and move on to other disciplines to study, you will look back with satisfaction on what you've achieved. And you should do so.

Those of us who are truly educated in Anatomy realize that we have barely scratched the surface of what is known and unknown about Human Gross Anatomy. One of the values of a course like this is to make all of us aware of the immensity of what is known. As you advance in your education, you will encounter individuals who have spent their lives studying one infinitesimal recondite process in the workings of the body, you will be awed by the length and breadth of their knowledge, and you will be humbled by your own lack of awareness. And that humility is the beginning of wisdom.

CHAPTER 10A:
Discourse with the Donor

"Don't be ridiculous. A student removed some gold bridgework from a cadaver's mouth, and took it home?"

"It's true. She did. That's definitely against the rules!" Donor Body 2436, Table 6 had arrived at my bedside again, this time with bad tidings.

I was shocked. "It is. And I don't have to tell you it's also against the *law*."

She took the topic one step further. "It's not just against the law, it's a *felony*."

"See? As I said, I don't have to tell you."

She smiled, and came a little closer to my bedside. "I thought all that felony business only applied to body parts."

"The dental restorations of the patient are considered body parts, and desecration or larceny of that part is the same as any other."

"I suppose that accounts for the fact that you carefully keep the lab door locked when no one's around. The students can only get in by swiping their student ID cards. Otherwise, I suppose, anybody passing by could be tempted to steal the gold from our mouths." She was speaking of the cadaver population as if it were a community.

"Yes, and I also keep that door to this bedroom locked, but that doesn't prevent some people… *like you* from coming in and disturbing my sleep."

"Yeah. I'm not like all those other women who crave to come into your bedroom," she said mockingly. "You've noticed the locked door can't keep *me* out."

"Yep," I responded wistfully. "So many women. I'm such an Adonis. Last week a woman was pounding on my bedroom door for an hour, keeping me awake. She was hollering to me to open the door, but I remained resolute."

"Seriously?"

"Yes. Then I finally got up and let her *out*." I was pleased with myself, and my little joke.

She shook her head in disgust. "A person can only be young once, but you can be immature forever." Then she chuckled. "Reminds me of Jane and Charles. They spend so much time in the lab alone together, they're positively zygotic."

"Jane and Charles? I had no idea."

"They purposely plan their rendezvous at times when it's likely no one else is in the lab. They're crazy about it."

"They're crazy about the lab?"

"The rendezvous, silly. They cloister 'way over at Table 35 and who knows what they're doing."

"Quizzing each other on the branches of the brachial plexus, I bet." I chuckled as I thought about my student days in the Anatomy Lab. It was the closest thing to a date that my new wife and I could afford, in those days.

"Yep, no doubt." She leaned over my bed. "Hey, professor: we were talking about felonies taking place in your Anatomy Lab. What are you going to do about it?

"I'm not sure…"

"You allow students to come in at any time, day or night, and they do. Professional students keep late hours, and some of them don't sleep much, I've noticed. Some of them stay past midnight, especially during Reading Week."

I pondered the wisdom of 24/7 access to the cadaver lab. "And there are some who sleep *too* much. They're sleeping when they should be studying their Gross Anatomy."

"Ha! They sleep during your lectures!"

I quickly changed the subject. "We're getting four more donor bodies tomorrow night, so you'll have even more members of your little community in the lab."

She smiled. "I remember those two spiffy-looking guys that brought me from the morgue. Such a relief to be out of that place. Being stacked in there wrapped tight in a shroud, like a mummy." She suddenly stopped speaking, and pondered that for a few seconds. "Those hearse drivers do work odd hours."

"Yes, they generally come in at night, to escape the attention of anybody who might be lingering about."

Her smile had become a grin. "What a time they had with me. The front wheels of their gurney were stuck in the "left" position, and the wheels kept yanking the gurney out of line." She paused, remembering. "We kept whacking into the door frames, even on the service elevator. They would reverse and try again, but the gurney continued to swerve to the left. It was like I was refusing to come into the lab."

"Maybe you were having a premonition about what goes on in the lab.'

"Oh, I knew what goes on in the lab. I had no problem with it," she said, pausing. She rolled her skeletal fingers into a fist, and rested her mandible on it, and smiled at me. "I actually enjoyed annoying those guys. They're so *serious*!"

"I'm not surprised that you caused trouble for them," I said, teasing her. I felt compelled to speak up for my friends the undertakers. "Normally, their equipment is first-rate. Those are the people who deliver bodies for funeral homes. They bring the bodies in a hearse, and they dress accordingly. That's why you called them "spiffy." Personally, I think it's important to dress up when transporting human bodies, and to handle them with dignity."

"Even when they work late at night," she said.

"They have to work late at night. When you think about it, can you imagine a patient or a family coming across a body bag on a gurney *in a medical school?* Talk about bad publicity for the school." I suddenly realized that remark might offend her, and I was loath to do that. "I'm just saying, it's not the kind of image the medical school wants to present to patients."

"No, I suppose not," she said.

"Furthermore, the Dean would have a hissy. We could kiss the cadaver lab goodbye, immediately. We'd be teaching human anatomy by drawing body parts with crayons. We'd be doing "virtual dissections" on computers in the library. And instead of human cadavers, we'd be dissecting animal parts obtained from a slaughterhouse."

"Ha!" she said. "Cow's eyes and pig hearts, from the slaughterhouse. Nice."

"Cat pancreas! Bovine testicles!" I agreed. "Actually, that's not so crazy. The Bursar once seriously proposed that we dissect a cat, and extrapolate the cat anatomy to human anatomy. He'd heard at an Administrators' Conference that some schools teach Anatomy that way. It's like learning to repair a car by taking a bicycle apart."

She raised her nail-polished fist to an imaginary Bursar. "Here, Mr. Money Bags! Extrapolate this!" She began chuckling, then laughing, then she guffawed so hard that she placed her hand where her abdomen used to be. With her other arm, she leaned on the bedpost as she shrieked, "Tripe intestines! Earthworm livers!"

Her one risorius muscle was working overtime, while the contralateral side of her face drooped, like a person with Bell's palsy. She lost her balance; I sprang from my bed as suddenly as a cricket, and reached out to catch her. With my gentle help, she regained her posture, and her demeanor.

She turned her attention to her appearance. Her skin was the color of parchment. She reached down with one hand and straightened up her flopping abdominal obliques. It was quiet in my bedroom.

Trying to regain my own composure, I said seriously, "To be honest, I don't think some professional schools would object to that kind of anatomy lab. They'd be grateful that we're saving money. Most school administrators don't know much about the anatomy, or the lab, and they don't care to know. They've never even been in my lab while a class was in session."

"Never? Not once?" She obviously took this offense very personally. "That would make them pretty ignorant about a major part of the curriculum in the school."

"And ignorant about Anatomy in general. At our last faculty meeting someone referred to the cadavers as 'corpus delicti.' He meant "corpses."

"Corpus delicti," she said, musing. "I think that's a legal term for information about a crime." The former English professor wore a condescending smile. "Speaking of crime, you need to deal with the bridgework larceny going on in the lab."

"We have security cameras in the lab," I said, grimly. "I'll see what I can come up with. I just have to decide if this is a mountain I want to die on."

"Your little felon hasn't taken *my* bridgework, at least not yet. I hope you 'come up with' something before she does." She nodded in agreement with her own statement. "Are you going to follow up on this?"

"I'll see what I can do," I said.

"I rather saw that as a yes or no question."

"Then yes," I said, glancing at the clock on the nightstand. When I looked back, she'd disappeared.

CHAPTER 11:
How to Take Practical Examinations

That Special Time

Most professional school faculty members enjoy teaching, and we especially look forward to the week before a practical examination. We know, even better than our students do, that those students will soon be in other classes where they'll need to know Gross Anatomy, and we are therefore motivated to present it to them. For their part, seeing the practical examination rising on the horizon like the sun in the East, the students' attention is on tip-toe. *Nobody's blowing the faculty off this week.*

For those of us who love Gross Anatomy, the practical examination is a special time on the calendar. For the student, it's the culmination of weeks of study. It's a chance to show off and display his or her knowledge, and the application of that knowledge to an actual human body. For the novitiate, practical examination preparation is by definition a time of rigorous reflection (and reflection, and reflection again).

After the examination, it's customary to meet your colleagues for a time of further reflection, sometimes over an adult beverage. We call such a meeting "The Post-Gross Toast." My parting shot at the students as they leave the examination and turn in their answer sheets is the same advice I implore you to follow: "Please party responsibly!"

Protocols of Most Practical Examinations

Let's start with this: in a professional school context, it's considered gauche to cram during the last few minutes at the door of a practical examination. Your learning should be completed before you show up. When you get to the examination, you'll find that the students congregate outside the door of the laboratory, waiting to be admitted to the laboratory *en masse*. Soon the hallway is packed with people, and, since it's universally considered bad form to study at this point, they mill around like theater-goers between acts. Don't be the guy crouched in the corner muttering over an atlas. Be maximally prepared upon your arrival, so last-minute panic is unnecessary. The chances are good that material you're cramming at that point won't be on the exam anyway.

Once you go into the laboratory, you'll be directed to a particular station where you'll begin. Each student's beginning station is assigned randomly in an effort to keep potential collaborators separated. When you arrive at your beginning station, most schools allow a minute or two for students to make notes on the back of the answer sheet. These notes are usually mnemonics or drawings of a plexus, like the brachial plexus. Or a sketch of the branches of the internal iliac artery. Things like that. For two minutes, you'll see classmates scribbling like a waitress at a table for twelve. Since it's likely there won't be enough time for drawing and listing after the examination starts, this time is allowed at the beginning.

Practical examinations involve a number of stations (usually between fifty and seventy-five), and students move like wedding guests in a receiving line, step-by-step through those stations. A fixed amount of time (at our school it's 60 seconds) is allotted to each station. A few of the stations are rest stations scattered along the way, at which students can pause and reflect and catch their breath.

Some particularly compassionate programs will post, at each rest station, a sheet listing the questions that are written on all the stations. The student at the rest station can therefore pause to review the questions (and, if the examination is multiple choice format, the answer options). At the rest station, the student has a little time to think, and to make more thoughtful decisions.

For recording your answers, some schools use fill-in-the-blank answer sheets, while other schools use multiple choice questions. Of course, grading the examination is necessarily different in the two formats. Grading fill-in-the-blank formats is time-consuming and difficult for the faculty (and cannot be delegated to staff), while multiple-choice exams can be graded electronically, saving the professors a great deal of time. Furthermore, on multiple choice tests, it's easier to collect and analyze data, like which questions were missed, and which distractors were incorrectly selected. This data can be helpful in identifying shoddy questions, and in improving the examination next year.

What's On the Practical Examination?

Naturally, the students are anxious to know what they'll be held responsible for, on the upcoming examination. At my school, here's how they know:

At each laboratory session, there is a written Laboratory Exercise, comprised of pages in the Laboratory Manual, or a handout peculiar to your course. Each Exercise includes a *Structures List for the Individual Laboratory Session*, listing the structures that the student should learn during that session. The week before the practical examination, all the *Structures Lists for the Individual Laboratory Session* lists are compiled into a single *Structures List for The Examination*. It's simply a melding of all the *Structures Lists for the Individual Laboratory Sessions*. Instructors will print and post that *Structures List for the Examination* a week prior to the examination. You can understand that the *Structures List for The Examination* contains all the structures the student has pursued in lab, and therefore all the structures the student will be held responsible for on that particular practical examination.

The structures on any such lists are usually listed in categories: nerves, blood vessels, muscles, and so forth. In each category the list is in alphabetical order. Shortly before the midterm examination, the *Structures List for The Examination* is posted for the students. At the examination, each student will be given a clean *Structures List for The Examination* to carry with them on the examination itself.

The students receive this Structures List a week prior to the test, and they use it as a study guide in their preparation. Then, while taking the practical examination, they carry a clean copy of that structures list with them. It's a good system; it eliminates problems with the most egregious spelling. (Even with the list in their hands, some students misspall wrods on the tests!)

The problem I find with this system is that during the examination, an unprepared student spends entirely too much time sorting through the structures list, when it would be more profitable for him or her to spend the time looking carefully at the structure itself, and the surrounding structures.

Each practical examination question typically takes one of two formats: it can be strictly "identify this structure," or it can be a question about the structure, for example: "Which nerve innervates this structure?"

Here's the thing: if I tag the long head of biceps femoris, and ask you which nerve innervates this structure, "long head of biceps femoris" needs to be on the *Structures List for the Examination*, and so does "tibial nerve." If "tibial nerve" is not on that list, is it a fair question? I don't think so.

Of course, words like "true" and "false," "yes" and "no," and the like are assumed to be O.K., despite their absence on the structures list. They are not anatomical structures.

Many schools incorporate other anatomical disciplines in their Gross Anatomy practical examinations. Students taking the Gross Anatomy Practical Examination may be advised that their current classes in Histology, Embryology, and Radiology may have submitted questions for the examination. Professors in those disciplines sometimes place a few practical examination questions on the Gross Exam.

Specific Protocols for Practical Examinations at My School

Here are my last-minute reminders, given to the students, regarding protocols on the Practical Examination in Head & Neck. I believe these apply to most practical examinations, at most institutions. You may want to check these out, ahead of time, with your faculty before the exam:

Nerves:

1. Regarding cranial nerves, you can either write "CN" followed by a roman numeral, or the name of the nerve.
   Example: "hypoglossal nerve," or "CN XII." The answer "CN 12" in incorrect.
2. For innervation of the muscles, if there is a named terminal branch for the muscle, write that. If not, the cranial nerve will suffice.
   Example: If cricothyroid muscle is tagged, and you are asked its innervation, you should write "external laryngeal nerve." "Vagus nerve" would be incorrect. If a muscle of the tongue it tagged for its innervation, you may write "hypoglossal nerve" or "CN XII."
3. If a specific nerve is tagged, write the name of the tagged nerve.
   Example: If the chorda tympani branch of facial nerve is tagged, please write "chorda tympani." Writing "facial nerve," or "a branch of facial nerve," in this case, is incorrect.

Ganglion vs Chain:

If the tag is on the ganglion write "ganglion." If the tag is between ganglia, write "sympathetic chain."

Abbreviations:
Only the following abbreviations will be accepted on the Practical Examination: a., v., m., n., lig., CN., L and R.
All other abbreviations, including, but not limited to: SCM, TMJ, and IJV are not acceptable.

Right/Left:
You don't have to include "right" or "left" in your answer, unless you are asked to do so.

Punctuation:
Punctuation is important on your exam. If the nerve is the McBurney point, you really shouldn't write "McBurney's point." To emphasize the concept, the story is told of the woman stating, "Those dead bodies in the cadaver lab are my ex-husband's."

Radiological images:
All radiological images will be given to you in their appropriate, anatomical view with orientation. None of the radiographs are upside-down, or mirror image*.

*Parenthetically, in my class at the dental school, I was able to obtain a radiograph of a cadaver wearing full upper and lower dentures that were made with porcelain teeth. Obviously, there are no roots on denture teeth. Only the crowns (tops) of the teeth are present, and if they're porcelain they show up on the radiograph. For a unique challenge, it was displayed as an annually recurring joke question on the practical examination. What's the disease this patient is suffering with, that results in the loss of the roots of all the teeth? Usually, the student assumes, incorrectly, they are natural teeth. *It's a joke.*

On "fill-in-the-blank answer sheets, how specific should my answers be?

1. The questions that we put on the lab exam will be specific and you must be specific when you answer them. The answer sheets that your professor provides have small boxes for you to write answers. If you are writing something that doesn't fit in the box, you are probably answering incorrectly. You may be describing the structure instead of naming it.

2. If you are writing a structure's name that did not appear on the Structures List for the Examination, you are answering it incorrectly.
3. Be careful when writing notes to yourself, or otherwise defacing the answer sheet. The answer in each box is graded *by hand* by a faculty member. If there is writing, or a circle, or check marks, make sure it cannot be mistaken for an incorrect response when the exam is graded.
4. Some students like to draw a plexus, or write a mnemonic, or list other reminder devices that they can use on the examination. Two minutes will be allotted at the beginning of the examination for you to write them *on the back of the answer sheet.* Please don't write them on the front. (See #3, above.)

Practical examinations are intentionally designed to be pressure cookers. The limited amount of time at each station places a burden on some students at some stations. Furthermore, as the semester goes along, and as the cadavers are being dissected more and more, it becomes increasingly difficult and time-consuming to orient oneself while identifying the tagged structure. If you are confronted with a structure you can't identify, it's always a good idea to find a structure that you *can* identify, and work from there.

Horses and Zebras

When taking a practical examination in Human Anatomy, try not to second-guess the question. Usually the question is straightforward, and should be handled as such. *Don't assume the professors are trying to trick you*. Your first impulse answer is correct in most cases.

It has been my personal experience that if you go back and change answers later, sometimes you are "overthinking" the question. Remember: the professors are testing your knowledge of the Anatomy; they are not testing your detective expertise. If you see a nerve within the carotid sheath, you know it's vagus nerve; don't use convoluted mental gymnastics to convince yourself that it's an aberrant external laryngeal nerve. There's a saying regarding this problem with Anatomy examinations: "If you hear hoofbeats, look for horses, not zebras." You'll break your professor's heart when he or she sees that you wrote the correct answer first, and then crossed it out and changed it to an incorrect one. Just answer the question. For more about this, see "Changing Your Answers" page 239.

Setting Up The Exam

At my school, on the last regular laboratory session before an examination, the professors go from table to table, asking the students which dissected structures on their cadaver are exemplary. The instructor makes an "excellent" list, mentally or on paper: "table 7 has an excellent example of chorda tympani nerve." When setting up the exam, the instructor can use these already-dissected structures. The instructors may not use all of the structures on their "excellent" list, but when they do, this practice saves them valuable time. You can help your professors by pointing out structures that you're proud of: a sphenopalatine foramen, or a trochlear nerve, for examples.

Generally, the practical examination takes two days to set up. The first day, members of the laboratory faculty spend several hours identifying and tagging the structures that the Course Director wants tagged. This is agreeable work for the lab instructors, most of whom like dissecting. At my school, as at most schools, we try to find the best representative example of a particular structure, and we work assiduously to remove or reflect enough other tissue to eliminate distractions and make the question a fair one.

For several hours, the laboratory gets very quiet, as we dissectors are concentrating on the work at hand. Occasionally, one of us may spend a long time dissecting a certain structure, only to eventually break it, (which may lead to the use of some adult language), or the instructor may find that the cadaver has an anomalous or otherwise unacceptable example of the structure we're looking for. Difficult dissections can be as frustrating for the instructor as they are for students. In that case, we may go to the other side of the cadaver, or move to a different body.

You've probably thought of this already: it's important for the professors to "shuffle" the cadaver *tables* for the examination, so the students will be less likely to associate a particular structure they themselves have dissected, with their particular table.

There are at least two different protocols by which the examination can be set up:

1. Presently at my school, we make six columns on the whiteboard at the front of the room, each with a heading corresponding with one of the six laboratory sessions for this unit. As we prepare the examination, each professor completes a dissection, making it a practical exam station. Then he or she writes the name of the structure on the whiteboard, in the appropriate laboratory session column. We strive for an

equal number of stations (practical exam questions) to be derived from each laboratory session the class has completed.

2. At my former school, the Laboratory Examination was in multiple choice format. (In this format, the students did not carry Structures Lists with them.) The Course Director takes questions from a Test Bank, writes each multiple-choice question on a 3 x 5 card ahead of time, and puts all the 3 x 5 cards on the windowsill. When the professors show up to prepare the exam, we start by taking one of the 3 x 5 cards, and we dissect out that structure on one of the cadavers. When that dissection is completed, the instructor drops the 3 x 5 card at that station and picks up another one from the windowsill.

There are, of course, a great number of variations on these basic themes, and some Course Directors are more creative than others in the construction of their practical examinations.

Number 1. above includes the added rigor of requiring the student to come up with the answer entirely on his own, but it also leads to questions about misspalling, especially if students are not allowed to carry a structures list for the Examination. How bad does my spelling have to be, in order to be incorrect? #2 is much, much easier to grade, and in fact a secretary can do it for you. There are pros and cons each way.

Once all the structures are tagged, the secretaries begin setting up and printing the actual questions associated with the stations. There is great potential for confusion on practical exams, and therefore extra care is taken in their preparation. In addition to the cadaver stations, some stations consist of bones or skeletons, some are prosections, and some even have diagrams and photographs to test the students' knowledge. At our school, there's often an embryology plastic model, or a pathological specimen, or radiographic images of various sorts, mixed in with the cadaver questions. And practical examinations offer an excellent opportunity to use plastinated specimens.

Sometimes on practical exams, we make one or more "joke" stations, in an attempt to break the tension that students feel while taking the test. The joke stations are designed to be funny, and they cannot be graded incorrect. For example, a few years ago, we took three blocks of wood and arranged them in a triangle, and asked the students to identify it. One of the multiple choice options was "lumbar (lumber) triangle." ("Lumbar triangle" was on the Examination Structures List.) Inexplicably, only a few of the test-takers found it funny.

Once we tagged the foramen of Winslow, and then we listed, along with "foramen of Winslow" answer options designed to be funny: foramen of Smith, foramen of El-Sharawy, and foramen of Forbes. (Smith, El-Sharawy and I are instructors in the course.)

After the examination is administered, a secretary or the diener takes one or two detailed photographs of each station, to aid in defense of the question, should a student challenge it later.

There are instances when a student thinks a question is unclear, or ambiguous, or unfair. An example of "unfair" is: the material tested was not part of the assigned dissections. Another example of "unfair" is an unclear pin placement. Each class will have its own protocols for dealing with students challenging the legitimacy of a question on the practical examination. For examples:

1. In one protocol, a complaint sheet is included with the blank answer sheet that's distribute to the students at the beginning of the Exam. The student writes their complaint, and inserts it into a box provided as they exit. Later, together, the professors address each complaint, armed with the meticulous photographs of the stations, and they decide if the complaint was warranted. I must tell you: I always felt the committees judging the complaints were more than fair. We *do* want you to succeed, you know.

2. In the fill-in-the-answer-entirely-from-memory type exam, the student will know that a professor grades these examinations by hand, and therefore can write complaints in the margin of the Exam, secure in the knowledge that the complaint will be noticed, read, and ratiocinated, during the grading of the examination.

3. In another, less ideal protocol, the student is encouraged to call one of the invigilators over to the station *during the exam*, to voice his or her displeasure. This is the *worst* protocol for dealing with the problem of unclear, ambiguous or unfair questions. It can be contentious in the middle of the exam and it poses a distraction to the other students. And it employs only one faculty member, in a brief time frame, to reach a conclusion about your complaint. Other faculty may disagree.

This is not to discourage you from seeking redress for a bad situation that could affect your grade. Certainly, if the question is unclear, if the pin is obscured, if there's tissue blocking the structure being asked, the student is well within his or her rights to call the instructor over to rectify the situation. That's what we're there for.

One more thing should be said about this: I have known instances when a student intentionally obscured the question, with the intent that other students would miss it. I know of at least one instance when a student moved the pin. Be aware that the potential for this misbehavior is the reason we have students begin at assigned stations. We know the sequence of students passing by the station. Professors faced with the question "who moved the pin?" can do a little detective work and figure out the point when everybody started missing the question. That leads to the person who moved the pin. It's pretty easy, really. Don't let the first student to miss that question be the person who followed *you*.

One more thing should be said regarding complaints about a practical examination question. It has been my experience that if the information you present is dispositive, the professor will be more than happy to improve your grade. So don't hesitate to bring errors in the exam to our attention.

Make a Mocktical Exam

At my school, the students put together their own practical examination, a few days prior to the actual practical examination. Since the students' examination is a mock practical examination, they call it The Mocktical Exam. I think this is the best final-stages preparation for a practical exam. It should be undertaken shortly before the actual practical examination, after most students have studied the material well, but it should also allow time for remediation of "gaps in knowledge" that a student discovers when taking the Mocktical Exam. Typically, the Mocktical is on a Friday, and the actual examination is on Monday. The intervening period is a time when I, and the other professors, receive several emails covering newly-discovered lacunae in knowledge that various students have discovered on the Mocktical. And we are happy to address any and all such emails.

In preparing the Mocktical Exam, the team at each dissection table tags two structures (could be more), and writes questions associated with those structures. The Mocktical Examination is left in place for several days, and students can come to the lab and take it whenever they want, as often as they want. Groups of students can collaborate on their answers. Our laboratory is available to students 24/7 for students to review. Students should work hard at keeping a light attitude about the Mocktical, not complaining about the questions, nor arguing about which answer is correct. *Dude: they're not graded.*

Usually the Mocktical Exam is given in the same protocol as the actual Practical Exam: is it multiple choice, or fill in the blank? Are students allowed to carry a Structures List? During the Mocktical Exam, students are usually allowed as much time as they need at each station. If it's multiple choice, the strict protocols may not be followed. For example, on an actual examination at my school, all possible options on multiple choice questions are put in alphabetical order. On the Mocktical Exam, they likely will not be.

The purpose of this examination is to point out to the student areas where he or she needs to do more work. There should still be a day or two (typically a weekend) after the Mocktical and before the Practical Examination, during which the students can address those deficiencies.

It has been my experience that the students' Mocktical Exam is typically more difficult than the actual Practical Examination, and I suppose that's a good thing.

Solving the Problem of the Difficult Practical Examination Question

In the course of taking a practical examination, sometimes you will come to a station, and have no idea what the structure is. As an Anatomy instructor, I myself get this all the time: a group of students will call me over to their dissection. Their dissection looks like a trailer park after a tornado. They pick up a string of tissue, yank it in my direction and ask, "What's this?" Very often, at that first moment, I have no idea what it is, myself. When that happens, I usually follow a certain sequence. Students might think about employing the same protocol. To wit:

Forbes' Method of Practical Anatomy: If you are confronted with a structure you can't identify, find something that you can identify, and work from there. The key to solving a practical examination question is to locate *first* a structure in the immediate area that I *can* identify. An *easy* structure. Working from that known structure, toward the one in question, I can usually gain some clues.

Example: Identify the nerves at the asterisk.

The sciatic nerve is the biggest nerve in the body, and its superiormost extent in the thigh is right in the middle of the backside of the buttocks. Take a look at *Netter 5th ed.* 490, entitled "Arteries and Nerves of Thigh: Posterior View," and notice how the big, yellow sciatic nerve runs directly down the middle of the thigh. Sciatic nerve is impossible to mistake because of its distinctive size, and it runs straight "north-south." We use that nerve as the landmark to identify the structures that surround it. We know that sciatic nerve emerges from the pelvis just inferior to piriformis muscle, and at the same place, inferior gluteal nerves and blood vessels also emerge (where the asterisk is). There's my answer.

Here's the point: working from sciatic nerve, we can proceed to identify many of the structures in the gluteal region. This technique is a matter of finding something you *do* know, and then moving from there to the structure that you *don't* know.

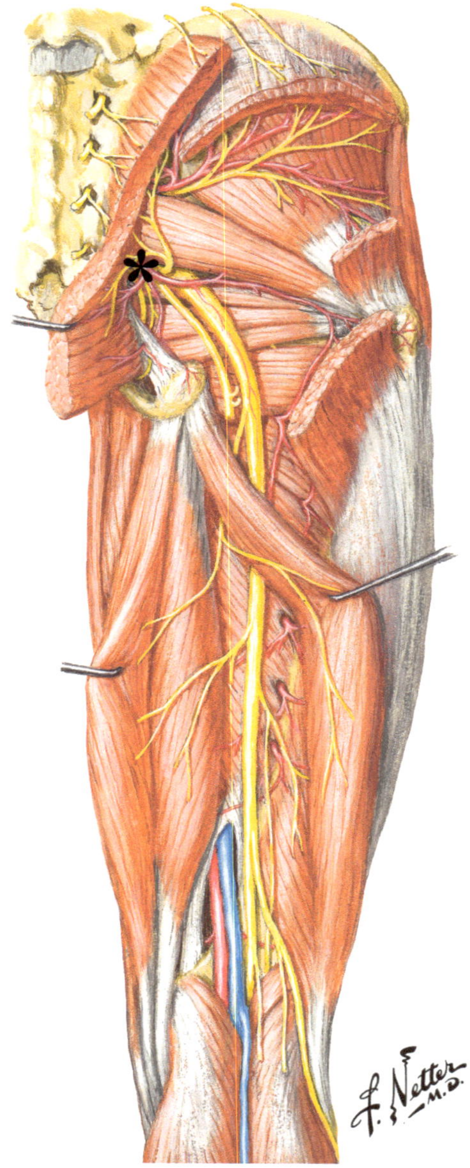

Netter Atlas of Human Anatomy, 5th Edition 490.

Know Your Instructor

The student should be aware of this important fact: each instructor has his or her own emphases in teaching the unit. This is true of all disciplines, not just Anatomy. These areas of emphasis naturally show up on the examination. Therefore, the student is wise to spend some time with *each* instructor in the lab, taking written and mental notes on what is being taught, and how much emphasis is being placed on a particular structure. For example:

Example: Identify the nerve and the artery.

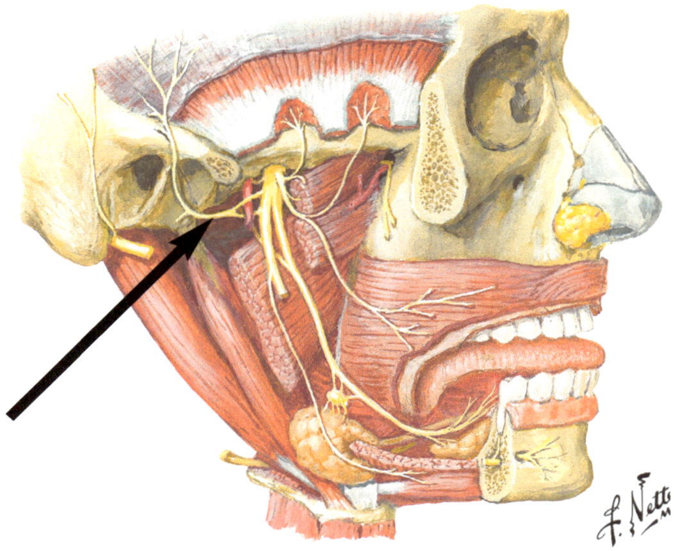

Netter Atlas of Human Anatomy, 5th Edition 45.
Copyright 2011 Elsevier Inc
All rights reserved. www.netterimages.com 10791

Notice the auriculotemporal nerve at the end of the arrow. As it passes horizontally across the infratemporal fossa, the nerve splits right down the middle for a short distance. Passing through this split in the nerve, like thread through the eye of a needle, is an important artery: the middle meningeal artery. After splitting to allow the artery to pass through, the nerve *rejoins itself*, and continues on posteriorly as a single strand. I don't know of another place in the body where such a thing occurs. I find myself wondering *how it happens*, exactly, as the body develops. Considering that there's fluid running through the artery and an impulse in the nerve, this situation is like a water pipe penetrating the middle of an electrical wire. It is strange indeed, but it is clearly present in almost every cadaver.

Since I marvel at it so much, I always mention this relationship in the lab, I always try to dissect a good example of it, *and I invariably put one on the practical exam.* I think most instructors follow that pattern. So I encourage you to work with *each* instructor, and make notes (written or mental) about what that instructor finds interesting, both in lecture and especially in Anatomy Lab. One instructor may be hung up on the chorda tympani nerve, another on the submandibular ganglion, another on deviated nasal septa. The things the instructor discusses and points out are the things that are important to him, or her. Those things will naturally show up on the examination.

The "Right Side" Rule

Drawings in an atlas almost always portray the subject's right side. That's the way you'll burn them into your visual cortex. When you come to a practical exam question on the cadaver's left side, you'll need to make that adjustment: you'll need to think in terms of a mirror image of what's in your brain. But that can be done, and it's the best way to approach learning Anatomy. For example: when you envision the brachioradialis muscle (whose name tells you that it runs from the brachium to the radius), its configuration is a little different from most of the other muscles of the forearm. You'll see it on the person's *right* side on Dr. Netter's plates, but there is a 50/50 chance that you'll see it on the *left* side on a practical examination. Nonetheless, the practice of burning the image of that (right side) brachioradialis into your brain is *very* helpful. When you're standing at a station on the practical examination, if you see the cadaver's left side, you'll work to recall an image, and you'll need to flip it over in your mind to *the left side*.

Confused? When considering the left side and the right side, remember Yogi Berra, the famous baseball catcher for the New York Yankees. He said that batters who can bat from either side of home plate are amphibious. You need to be similarly amphibious in your preparation for the practical examinations.

Identifying Arteries

In anatomy atlas drawings and in most textbooks, there are branching patterns for the arteries. (Not so often for the veins, because they are highly variable and highly unreliable). Let's take facial artery, for instance. Most textbooks will tell you that facial artery branches from external carotid artery just above where the lingual artery branches. From there, facial artery loops under the angle of the mandible, under the bare area and diagonally across the face (You can see this on *Netter 5th ed.* 69, entitled "Arteries of Oral and Pharyngeal Regions.") You will need to memorize that material for your written exam, but when you do your dissection, most of the time you will not find this tidy arrangement. In fact, very often, as much as forty percent of the time, facial artery branches off from external carotid artery *together with lingual artery as a single artery*, then the two arteries separate. Patterns can differ from what you see in your textbooks and atlases. How can you be expected to identify facial artery in this instance? Here's the secret: most arteries are more easily named by *where they go*, instead of their branching pattern.

This technique for identifying arteries is especially important inside the pelvis, where branching patterns of blood vessels are notoriously variable.

N.B.: William of Occam (also spelled "Ockham") should be a big hero in healthcare schools. The principle he is most known for is Occam's Razor: "*Among competing hypotheses, the one with the fewest assumptions should be selected.*" Whether you're identifying structures in the cadaver lab, or reaching a diagnosis for your patient, it's wise to apply Occam's Razor (also known as "The Law of Parsimony"). In the discipline of Philosophy, a "razor" is a principle that allows one to rule out ("shave off") the unlikely explanations for a phenomenon.

A practical examination is an excellent place to apply Occam's Razor: the best explanation of something is the simplest, using the fewest assumptions. When attempting to identify an artery in the pelvis, think first about *where it terminates.* I'm not recommending that you ignore completely the branching pattern. I'm recommending that you de-emphasize the branching pattern, and focus on the destination of the blood vessel. That will almost always give you your answer. If you have to do all sorts of mental gymnastics to convince yourself that the branching pattern makes it something else, you're probably wrong.

Closely related to this advice is this: some students hold the view that it's the job of the professor to try to trick you into selecting the wrong answer. That's a puerile, sophomoric approach that is occasionally used by a teaching assistant. In graduate or professional schools, you'll find that the professors merely want to determine if you know the material. They are not assessing your cleverness, nor your test-taking ability.

Identifying Nerves

Here's a tip to keep in mind when taking practical examinations: if the nerve is motor, it must go to a muscle. If it's sensory, it bypasses the muscles.

Identifying Veins

For the most part, Anatomy courses don't require you to memorize many veins. There are some veins that are important enough, and reliable enough, that they will be part of the lesson. Great saphenous vein and cephalic vein are examples. But in my course, there are precious few veins that the students have to learn, mostly because I never met a vein that I liked. Veins for the most part are unreliable characters, like the *Les Miserables* characters Madame and Monsieur Thenardier, and, like those troublemakers, they're likely to show patterns of behavior you don't expect.

Cross Section Stations

Many schools will put cross section questions on their practical exams. It can be the *pons asinorum* for many an aspiring anatomist. The only way you can prepare for these is to study examples in textbooks. *Netter 5th ed.* 325-330 contains several; if your professor is stricken with the cross-section disease, you should prepare accordingly. The cross-sections in *Netter* are more than adequate for most Anatomy courses.

"Jump Questions"

This type of question has been made famous (notorious?) by Teaching Assistants in cadaver laboratories over the millennia. A "jump question" (also called a "second-order" or "third-order" question) is a question that forces you to follow several steps ("jumps"), in order to ultimately arrive at the correct answer. These are most commonly used for oral quizzing at the cadaver, allowing a group of students working together to come up with the answer. Although they are frowned upon as test questions, they actually can be fun when the group can combine their knowledge, and competition puts a little bit at stake. At my school there is a certain dark competitiveness among the TAs: "Who can design a question with the most arduous steps?" For examples:

Question: The Teaching Assistant points out coracobrachialis. The question is: What is the origin of the muscle that shares its insertion with the origin of this muscle?

Answer: Ribs 2-5.

****The key to answering jump questions is to work backwards.* ***

a. The structure pointed out is coracobrachialis.

b. Coracobrachialis' origin is coracoid process (as its name tells you).

c. The muscle whose insertion shares the coracoid process is pectoralis minor.

d. The origin of pectoralis minor is ribs 2-5.

Now try this one, which is more difficult:

Question: What is the action of the muscle whose origin is also the insertion of a muscle whose only action is opposing another muscle whose innervation is cervical spinal nerves and C3-C5.

Answer: Flex the arm at the shoulder.

****The key to answering jump questions is to work backwards.* ***

a. The muscle whose innervation is cervical spinal nerves and C3-C5 is levator scapulae.

b. Its action is to elevate the scapula.

c. The muscle whose only action is opposing levator scapula (depresses the scapula) is pectoralis minor.

d. The muscles whose origins are the same as the insertion of pectoralis minor are coracobrachialis and the short head of biceps brachii.

e. Both coracobrachialis and short head of biceps brachii flex the arm at the shoulder, and flex the forearm at the elbow.

f. The original question is: "What is the action…?" No matter which of the two muscles you pick, the action of coracobrachialis and short head of biceps brachii is the same: Flex the arm at the shoulder.

These sorts of questions present several problems:

1. Not all textbooks agree on exact actions for all the muscles in the body. Therefore, if you use a different Anatomy book from the one the instructor requires, you may learn the data a little differently. For example, if you use one of my favorite textbooks, Marieb's *Human Anatomy*, in the 4th edition it lists the actions of levator scapulae as "elevates, adducts scapula in concert with superior fibers of trapezius; tilts glenoid cavity downward; when scapula is fixed, flexes neck to the same side."[1] This is quite different from my Notespak: "elevates scapula," but the information is every bit as legitimate.

If your instructor uses jump questions, he or she will expect you to learn the material from a specific source: a Notespak, or a particular textbook. In my opinion, that limits your exposure to other sources, and diminishes your education significantly. Therefore, I myself eschew jump questions on examinations, but I enthusiastically promote them as learning exercises in the laboratory.

2. Jump questions require that you have specific data and memorize minutiae in order to get the question right. For example, the Anatomy course from which I derived the above questions distributes to students a Notespak that must be memorized completely, word for word, in order to be successful in the course. The abbreviated format in most Notespaks may not serve you well in a jump question.

3. Time is always an issue in jump questions. The instructor is asking you to answer several questions in rapid succession, rather than one straightforward question.

4. Jumps have a place in reviewing, and for grilling students in the lab. Often they are an opportunity for the Teaching Assistants to demonstrate how well *they* know the material, and that can be fun. However, on examinations, I want to determine if the student knows the material. I'm not particularly interested in swiftness.

Rest Stations

As described on page 202, *if you have a particularly kind professor*, he or she will set several sheets of paper at all the rest stations. On those sheets is a list of the questions asked at each examination station. Those sheets are provided so that you can review. If your practical examination is multiple-choice, the questions on the sheets will be complete with the answer options. This courtesy allows you to spend your time at the rest station more productively.

Some classes allow (encourage?) students to good-naturedly deface the "Rest Station" sign. For example, I've seen the sign altered from "Rest Station" to "Cardiac ArRest Station." Once I saw a stick drawing labeled "Two Students Walking a bRest Station." Anything in good taste that lightens the tension on a practical examination is salutary.

The Lightning Round

Many practical exams, including mine, will allow a "lightning round" at the end. Again, this is a courtesy found only in select practical exams. What's a lightning round? *After completing the usual practical exam*, where time at each station is typically somewhere between 60 and 80 seconds, the students will go through the entire exam again, with just 15 or 20 seconds at each station. This is a time to review your answers. Sometimes the professor will offer the lightning round as optional; sometimes it's mandatory. If it's optional you should *always* stay for it, and confirm your answers.

On the other hand, here is the worm in the "lightning round" apple: students sometimes change their answers from the correct answer to an incorrect one. It is very demoralizing for the professor as well as the student, to see that you have crossed out the correct answer and replaced it with an incorrect one. However, statistically speaking, if you're pretty sure you should change your answer, you should. See page 239.

Carrying a Structures List During the Practical Exam

As I've said, some programs allow the students to carry a clean copy of the *Structures List for the Examination* while taking the Practical Examination. If you're allowed to carry one, you should use the *Structures List for the Examination* like a drunk uses a lamppost: for support, and not for illumination. I often see the poorer students spending more time looking at the *Structures List for the Examination* than they spend looking at the specimen. Spend your time finding a structure you *do know*, and work from there to identify the tagged structure. Occasionally you will *almost* remember the name, but you'll not be able to bring it into your memory completely. *That* is the time to consult your Structures List.

Practical Examination Grading

Having taught at several universities, I have witnessed several strategies for administering and grading practical examinations. This is the approach to grading practical examinations that I like the best.

Immediately after the completion of the practical examination, a staff member takes careful photographs of each examination station, compiling a photo archive for reference when grading the students' answer sheets. Anatomy faculty are required to report to a "Grading Session" on the Tuesday morning following a Monday examination day. Practical examinations are hand-written, and thus require faculty (not staff) inspection for evaluation. At the "Grading Session," a subset of exam questions is assigned to each faculty member to assess. "Dr. Forbes, you will grade questions number 15 through 25." Dr. Forbes considers that particular subset of questions for all of the students, to assure consistent grading.

Because all the Anatomy faculty are in the room, the faculty member may consult with peers if issues of ambiguity arise. The photo archive of lab practical examination specimens is available for projection on a large screen, to help the faculty recall the dissection. At that time, if necessary, adjustments to the answer key can be made. It's like a basketball game when there's a questionable call; the officials will "huddle up" at the scorer's table, and look at slow-motion videos. Together, they reach the conclusion that your guy fouled the other guy. In the Anatomy Exam Grading Sessions, we professors huddle up and decide whether your answer did, or did not, commit a foul.

Once all the questions have been evaluated, they are double-checked by a second faculty member, and a grade is tallied on each answer sheet. Each answer sheet is then reviewed, and re-checked for mathematical accuracy. Once confirmed, the tests are sorted alphabetically, and collected by the course director(s) to record.

The logistics of running a practical examination are simply too arduous to be casual in the grading. Consistency is paramount: an answer marked incorrect or appropriate for partial credit on one answer sheet must be marked exactly the same in all the answer sheets. Therefore, a single faculty member grades a particular question for each and every student.

Avoid the Appearance of Evil

We Presbyterians consider the appearance of evil almost as iniquitous as evil itself. Many proctors of practical examinations, Presbyterian or not, feel the same way.

Therefore, when taking a practical examination, don't be looking around. Pay attention to the station where you are, and when you've completed the question, look at your answer sheet, or at the floor. *I know* that you're just being friendly when you signal your buddies that "This is easy!" but it's simply not fair to your classmates to do so. Like a joke in the TSA line, we will take your actions seriously. Furthermore, it makes your invigilators *uneasy about you personally*, giving extra suspicion and scrutiny to your demeanor and your actions, and that's never a good thing. When I see a test-taker looking around surreptitiously, like a man urinating in public, I become suspicious of that person. 'Nuff said.

Even the slightest indiscretion during an exam can get you in trouble. A Canadian person might innocently say, "Aye?" and a Spanish speaker might say as an encouragement, "Si!." Those could be misunderstood by the professors as the letters attached to options on the multiple choice exam.

For Examinations: Dress for Success

This might surprise you: I recommend you dress up for your exams, especially final exams. You should dress up when you take written exams as well as practical ones. I know, I know: it's popular among undergraduates to look like they've just fallen out of a boxcar when they show up for exams – to display the "I've been up all night" look.

I'm reminded of a friend of mine, who advised her adult children, "When your dinner guests arrive, greet them at the door dressed up, with a magazine in your hand. You want to give the impression that their visit is such a pleasure and so important to you that you were prepared long in advance, in fact you were so prepared that you were reading a magazine when they arrived." I think there's something to this. In every class I've taught, I encounter some students who dress up for exams. I think it's appropriate to do so. You don't need to step into the cadaver lab like a red-carpet movie star arriving at the Oscars. Just arrive well dressed. Eschew, for the moment, the baggy scrubs and out-at-the-knee blue jeans. After all, if you want to stand out among your classmates, why not stand out for your sartorial splendor? The practical examination is the culmination of an incredible amount of work, and, after the exam, that work will be completed. Why not celebrate that? You should be proud of what you've accomplished; let's dress accordingly.

Furthermore, I will tell you the truth: without fail, when a particular student shows up at the exam attractively turned out, the faculty notice it and remark positively about it. "Who is that student?" That's never a bad thing.

By the way, it is my considered opinion that professors should dress up for examinations also. And for ALL lectures as well. It shows respect, not only for the students, but for Gross Anatomy as a respected pursuit. For the faculty, a white coat works very nicely.

What Happens If I Come Late to the Practical Examination?

Resolving the dilemma of a tardy student is completely up to the Course Director. I will tell you that most professors have very little sense of humor about it. In my class, every year we have students who are entitled to more time on the Practical Examination for medical reasons, and the extra time for them was tacked on after the rest of the class had left the room. We allowed a latecomer *with a good excuse* to make up the few questions he or she had missed, during that time.

What's considered "a good excuse" in the preceding paragraph? Every professor will have his or her own standards, in this regard.

I am reminded of a student who came breathlessly to my study on the day *prior to* the examination. He had mistaken the date, and thought the exam was today. When I informed him that the exam was indeed the following day, he went on his way rejoicing, because he realized he had an extra 24 hours to prepare for the examination. When he took it, he achieved 100%. I'm thinking there's a lesson in there, somewhere.

What Happens If I Miss the Practical Examination?

Let's start with this: it takes around thirty hours of professors' time to set up a practical examination. Do you think your professors are going to set up a separate, new examination for you if you miss the Exam? Certainly not. The professors are not willing to use their valuable time to do an entire second examination just for you. And yet, if we use the original exam later for the absent student, that student can inquire from his or her friends what was on the examination, giving the absent student a significant advantage. Students are not unaware of this advantage, especially those who anticipate doing poorly on an upcoming examination.

Assuming it's an excused absence, the student takes a make-up examination. That make-up examination is necessarily *of a different format*. Typically, it will be a less desirable format than that employed on the original exam. This tends to discourage students from intentionally skipping the original one in order to gain more time preparing.

Therefore, in many schools, the protocols for a make-up examination are as follows: you get a list of structures to identify on your cadaver. You get ten minutes to look the list over and get your thoughts in order. Then you have one hour to identify the 45 structures, working at your own pace. Two instructors proctor. At the end of the sixty minutes, you will point out your structures, or answer questions pertaining to those structures, and the instructors each individually grade you on your answer. At the end of the session, the two instructors get together and settle on a final score for each structure.

Your professor realizes that there are legitimate reasons why you must miss an examination, like sickness. Of course, if you're genuinely sick, we don't want you to come and spread your pathogens through the entire school. We all understand that. Contrarily, sometimes a student describes symptoms that stretch credulity. Bill Murray's character in the movie *What About Bob?*[2] listed his signs and symptoms for Dr. Marvin, and they sound like a list I've heard before: "Dizzy spells, nausea, cold sweats, hot sweats, fever blisters, difficulty breathing, difficulty swallowing, blurred vision, involuntary trembling, dead hands, numb lips, fingernail sensitivity, pelvic discomfort... What if my heart stops beating? What if I'm looking for a bathroom, and I can't find it, and my bladder explodes?"

Believe me, if your bladder explodes, that is a legitimate excuse. Fingernail sensitivity? I'm not so sure.

While we're on the subject of pestilence, I have good news for you if you aspire to enter one of the health professions. I realize there are many motivations for doing so, but here is one that may surprise you: if you're in a healthcare profession in which you deal with patients, you're going to develop an enviable immune system. I spent the first twenty-five years of my professional life breathing directly into people's faces all day, *and I wasn't sick ten days in my dental career.* And that was before dentists began wearing masks. I've heard it said that kindergarten teachers are sick the entire first year of teaching, and then never again. The strong immune system you build as a healthcare provider is similar; it's a benefit not often mentioned in attracting people to the professions, but a noteworthy benefit it is.

What Happens If I Fail the Examination?

Each Anatomy course will have its own criteria for passing and failing an examination. Most institutions will expect you to achieve 70% in order to pass. Some schools will allow a re-test, with the same format as described in the "What Happens If I Miss the Practical Examination?" paragraph.

You may wonder why a failing grade on a practical examination isn't simply averaged in with your other grades in the course. In professional schools, the curriculum builds on itself to such an extent that most schools will not allow you to advance without proficiency in *each area* of Gross Anatomy. Big gaps in your knowledge of thorax anatomy will handicap you in your Cardiology clerkship, for example. Facility in all parts of anatomy is a necessary preamble to learning to be a competent healthcare provider.

What Happens If I Fail the Course?

If you fail the course, at my school you will have an opportunity to remediate, including repeating a Practical Examination. This is the case in most schools. The school may provide you a tutor (usually an upperclassman), and a cadaver to dissect, perhaps with other students who've failed. The cadaver may be left over from your course, but by that time most of the cadavers are wizened, and disarticulated to a great extent. So in many cases the cadaver is a new one. You'll take a week or two of your summer break to dissect the cadaver with others in your situation, with oversight by the tutor and the professors. At the end of that time, you will take a remediation practical exam and a remediation written exam.

If you fail again, you may be referred to a committee that deals with these issues at your school. There's more about the workings of Student Performance Review Committees and Academic Progress Committees on page 246 of this book. As a result of their deliberations, you may be offered a chance to repeat the year, or you may not. That is usually up to the Committee. The criteria for their deliberations are outside the scope of this book.

Format of a Mock Practical Examination (Mocktical Examination)

1. Tagged Structure: umbilical artery

Identify the tagged artery.

a. superior vesical

b. umbilical

c. obturator

d. inferior vesical

e. inferior rectal

Correct Answer: b

2. Tagged Structure: pelvic splanchnic nerve

Which type of nerve fibers are most likely to be found in the tagged structure?

a. pre-synaptic parasympathetic fibers

b. pre-synaptic sympathetic fibers

c. post-synaptic parasympathetic fibers

d. post-synaptic sympathetic fibers

e. somatic fibers

Correct Answer: a

3. Tagged Structure: Buck's deep fascia of the penis

Identify the tagged structure:

a. Colles' fascia

b. Dartos fascia

c. Buck's fascia

d. corpus spongiosum

e. perineal membrane

Correct Answer: c

4. Tagged Structure: skin of the inferior rectum below the pectinate line

Which lymph nodes does the tagged area drain lymphatic fluid into?

a. deep inguinal

b. superficial inguinal

c. internal iliac

d. lumbar

e. popliteal

Correct Answer: b

5. Tagged Structure: First webbing on dorsal aspect of foot

Which nerve innervates the tagged structure?

a. deep fibular nerve

b. superficial fibular nerve

c. tibial nerve

d. lateral plantar nerve

e. medial plantar nerve

Correct Answer: a

6. Tagged Structure: Abductor hallucis muscle
Identify the tagged structure:

a. flexor digitorum brevis muscle

b. abductor digiti minimi muscle

c. abductor hallucis muscle

d. flexor hallucis longus muscle

e. flexor digitorum longus muscle

Correct Answer: c

7. Tagged Structure: flexor hallucis brevis muscle

Which tendon splits the head of this muscle?

a. flexor digitorum longus

b. flexor hallucis longus

c. fibularis longus

d. flexor digitorum brevis

e. tibialis posterior

Correct Answer: b

8. Tagged Structure: Medial plantar nerve

Which muscle does the tagged nerve innervate?

a. abductor digiti minimi

b. quadratus plantae

c. dorsal interossei

d. plantar interossei

e. flexor hallucis brevis

Correct Answer: e

CHAPTER 11A: Discourse with the Donor

Her voice was inappropriately loud, like the voice of a person listening to loud music on headphones. "The pathology professor was in the lab yesterday. Some of the donors had tumors and internal lesions that they weren't even aware of in life. The pathologist was taking biopsies and tissue samples of disgusting diseases. It's amazing, some of the conditions they uncovered." She thought a moment. "Why do they care? The person is *dead*."

"They use the tissue for teaching and research purposes. Histological sections, and the like."

"She said my enlarged heart was common. *Common!* I object to that characterization. My heart condition resulted from a life of rigorous exercise, in a person with chronic pulmonary artery hypertension. I worked hard to remain fit, in life. I was a runner, for crying out loud. I ran everything I could, from fun-runs to marathons. I wasn't the least bit common."

"Not common at all. Still aren't. Now that I know you, I think you deserve a medal for what you've done. Donating your body to science is a courageous and uncommonly generous thing to do. I consider it heroic."

"Ha! A medal for distinguished service to the student 'body'! Get it? No, Habibi, it isn't anything like that. My body is a gift to the students; my intention is improving their lives and the lives of their patients. Besides," she said, looking down at herself, "if I earned a medal, where would I pin it?"

"I admire what you've done," I said, simply.

She ignored my remark. "The diener was tidying up the lab late last night, and…"

"The Dean tidies up the lab?"

"The *diener*, silly. He was putting things away, and one of the tables was missing their atlas. Who would sneak one of those smelly atlases from the lab?"

I said, "I've never understood that. You can buy a used older edition of the atlas on line for twenty-five bucks! Whoever took it, I bet their spouse or roommate was thrilled to have that stinky atlas on the kitchen table. It's bad enough that some students wear their laboratory scrubs home. They smell like formaldehyde, and many people hate that smell."

She sniffed her own underarms. "Hmmm… What a nice thing to say about me. And just when I thought I couldn't like you more."

"I'm sorry if I offended you. You know that I personally like the aroma of formaldehyde."

She nodded. I went on, "Those 'dirty' atlases smell like cooking bleu cheese. I don't know why they would steal one of those books. The school will require us to replace it with a brand new one, and they are very pricey."

"Is there valuable information in those laboratory atlases? Stuff they can't get somewhere else?"

I said, "The school presents each student with the atlas at Orientation. So they already should have one, at home. The only benefit to swiping one from the lab is to obtain a second copy, to avoid schlepping a heavy book back and forth from school. We encourage the students to use the atlases and dissection manuals casually, *in the laboratory*, even handling them with wet gloves if necessary, and laying them directly on the bodies. The atlases begin to smell, just like the bodies do, so nobody's tempted to steal them, as a rule."

"The bodies?"

"No. The atlases, of course.

"And the lab manuals? Aren't they supplied by the school?

"The students are required to *buy* a personal copy, and the school supplies a "dirty copy" at each dissection table."

"Are they helpful? Do they have study aids? Diagrams and cartoonish drawings?"

"The dissection manual is 'cookbook style' with detailed descriptions of steps in dissection. I suppose one could use it as a sort of "Cliff's Notes" version of the material."

My mind turned to something I'd wanted to ask her. "By the way, what was the kerfuffle at your table today?"

"One of my students discovered my maxillary torus, and thought I had a tumor. They got excited, and called the pathologist over."

"Oh, that's rich."

"I myself was surprised in life to discover that I had a torus palatinus. And now, since my craniotomy, I've discovered that I also have hyperostosis frontalis interna. Who knew?

I'm thankful that both of those conditions are benign." She paused. "I suppose there's no use concerning myself with them, at this stage."

"No," I said. We were quiet for a few moments, pondering the dime-sized bony dome in the middle of her hard palate. "Did the torus ever present problems for you?"

"Never," she said. "I myself never thought anything of it. I guess I thought everyone had a lump like that. My doctor told me the only problem would be if I needed dentures, which, as you can see, I never did."

For a few seconds we regarded the profound relief that patients experience when their doctor says their signs and symptoms are caused by something benign. She looked uneasy as she asked, "Will you donate your body to science?"

"No, of course not," I replied.

"Why on earth not?"

"I'm not dead," I said.

CHAPTER 12:
How to Take Written Tests

Go to Bed

Get a good night's rest before the test. It's only logical: if you're depending on your brain to bring up obscure things from the darkest recesses of your cerebrum, that brain needs to be well rested. After all your strenuous preparation, get some sleep before the examination. Set the alarm on your phone *and also* on an alarm clock.

In terms of being keen while taking an exam, caffeine will not necessarily avail you. The old joke "sleep is a symptom of caffeine deprivation" is simply that: a joke. Caffeine is not really a stimulant, you know. It merely blocks the effect of adenosine. Adenosine keeps in check the natural stimulants in your body. So caffeine is not the good guy; it merely blocks the bad guy.

Obviously, you'll do better on exams if you are (1.) adequately prepared, and if you are (2.) rested. Maybe your grandmother really *did* die this week, and honoring her memory compromised your study time. That's understandable. But if you felt lucky last night and therefore played poker until 2:00 A.M., you'll be taking the well-known approach *morituri te salutant* ("We who are about to die salute you."), as you perp-walk into the testing hall. Be assured: at that point, some of the power lines in your brain are down, and that factor is compromising your ability to think optimally. Instead, show up at your exams secure in the knowledge that you have prepared as well as your time and abilities can prepare you. The cognizance that you've done your best will help you relax.

Most animals are ruled by sunrise and sunset; they adjust their circadian rhythms, their "body clocks," to the light and darkness that are imposed on them by Mother Nature. Humans are the only species that overrides Mother Nature by lights of our own making, and by travelling to different time zones. We humans have the ability to stay up far into the night, and we do. Indeed, some of us routinely "pull an all-nighter," in preparation for an examination. One could argue that daylight is the natural time for wakefulness; the night's darkness is the natural time for sleeping. It may be risky to ignore nature's prompters in this regard.

The time you spend sleeping is time you're not spending studying, and when you stay up late studying, that's time when you're not sleeping. Since each person is an individual, how can we determine our "sweet spot," for maximum knowledge *and* maximum keenness at the time of the examination? It's a little like Goldilocks, who found the Papa Bear's porridge too hot, the Mamma Bear's porridge too cold, and Baby Bear's porridge *just right.* Your individual need for sleep should take preeminence in this decision. Don't deny your own experience. Stay up late studying, but get at least the minimum hours of sleep that has proven sufficient for you in the past. That's the Baby Bear's porridge.

The idea that one can continue to be keen after extended hours of being awake has been challenged in recent years, especially relative to physicians and residents in hospitals. In Jennifer Ackerman's excellent book, *Sex, Sleep, Eat, Drink, Dream*, she describes a study in 2004, in which a team of scientists were working in the Harvard Work Hours, Health, and Safety Study. The study reported that medical interns working thirty-hour shifts suffered twice as many failures of attention while working at night as interns whose work was limited to sixteen consecutive hours… "When people have been awake for 17 to 19 hours, their performance is equivalent to someone with a blood alcohol level of .05 percent," explains team member Charles Czeisler. When they're awake for twenty-four hours, it's .10 percent. "The risk of making a mistake after working for 24 hours is so great," he says, "that sleep experts and legislators in Massachusetts have suggested it may be ethically imperative for hospitals to notify patients if the doctor treating them has been awake for 22 of the last 24 hours."

Ackerman goes on to describe another conclusion by Czeisler. Studies of people who sleep only five or six hours a night on a regular basis (the average for most interns) have found that their median reaction time triples. "That means that when interns are driving home from an extended shift, if a kid dodges in front of their car," says Czeisler, "it will take them three times longer to move the steering wheel or put their foot on the brake." In 2005, Czeisler's group reported that interns working an extended shift of more than twenty-four hours had double the risk of having a car crash while driving home, and five times the risk of a near miss.[1]

I encourage you to give this some serious thought.

Study Old Exams

Take as many practice examinations before your examination as your time will allow. Studying practice test questions will shine a light on the areas you need to concentrate on, as you prepare for your actual examinations and quizzes.

Where can you get test questions to review? Many Anatomy textbooks contain examinations and case studies at the end of each chapter. Also available are whole books of test questions. One of those, *Board Review Series in Gross Anatomy*[2], now in its 9th edition, appears in the "Textbooks and Other Books" chapter of this book. Usually there are helpful explanations of the correct and/or incorrect answers to the questions, in such a book.

While most schools go to great pains to keep their test banks private, some schools actually make available their old examinations. Some fraternities and sororities maintain test banks for their members. And there are test questions on the internet, as well.

Is it easy? Is it fun? Definitely not. It can be extremely tedious. You must meet that tedium head-on with an iron discipline, and not yield an inch to it. You must have the tenacity of a horsefly on a marathon runner. If you can manage – aided by a combination of piteous entreaties to yourself, threats and bribes to yourself, and the occasional application of the lash – to drag your refractory attention span across the finish line by reviewing all the appropriate test questions, you will be better prepared for your own imminent examination.

Anxiety

You've heard it said that some nervousness before taking a test is a good thing. I tell my students that if they're not a little apprehensive, a little nervous, before a test, they probably won't do their best. My position has been that the nervousness keeps your focus on tiptoe, leading to the best possible outcome from your efforts. So if you're nervous, try not to let it bother you; it's probably a good thing.

On the other hand, I've seen nervousness overdone. I've seen students entering the examination room trembling, and vibrating like spin dryers. Obviously, the intensity of this nervousness has exceeded the beneficial stage. Bradley R. Staats, in his excellent book *Never Stop Learning*, states that "First, although some anxiety may be useful, too much is debilitating. We have difficulty making decisions or, possibly, doing much of anything. This is known as the Yerkes-Dodson Law." Dr. Staats goes on to say, "…anxiety and fear lead us to consider fewer options less carefully, and to remain wedded to our existing positions, even when we should change. So the other half of the Yerkes-Dodson Law is that although low levels of anxiety may improve performance, high levels worsen it."[3]

Should you be a little nervous as you anticipate a weighty examination? The answer, of course, is "of course." Usually, there's a great deal at stake. When you open the exam and get going, most of your anxiety will melt away like ice cubes in the sink, as you begin to concentrate and recall.

Answering Correctly If You Don't Know the Answer

If you're the kind of person who thinks that a "midterm" is a political election, and a "final" is the last game of March Madness, you may find yourself in difficulty in your Anatomy class. If you find yourself in a midterm or a final for which you are ill-prepared, try applying these tips. They might help you *a little bit*.

1. If you are sure that two of the options are correct, and if there is an option of "All of the above," then "All of the above" must be the right answer.
2. "None of the above" is frequently the wrong choice. This phrase is sometimes used by the professor as a "filler," if he or she doesn't have enough satisfactory distractors for the question.
3. The correct answer is more frequently the third option if there are five options offered to you. If there are only four options, it's most often the second or third choice. That's the way a professor will list them, in most cases. Obviously, this rule doesn't apply if the answer options are listed in alphabetical order, as they are at my school.
4. An option that is significantly longer or shorter than the rest is more frequently the correct answer. This is especially true if one option is *longer*; the test maker is trying to cover all possible student objections that could arise with the answer.
5. If you find that two options are exactly the same except for a word or two, one of those options is often the correct answer.
6. Don't forget to look for clues in other questions. That's one reason that you should *always* go back through the exam at least once.

7. Options that contain the words "all," "always," "never," or "none" tend to be incorrect answers. Options that include the words "most" or "usually," or "some" tend to be correct.

8. If the question and the option do not make grammatical sense when read as a complete sentence, the choice is wrong. If the correct answer is incorrect grammatically, you will have a good argument in Test Review that the answer as stated did not make sense.

9. Typographical errors usually do not appear in the correct answer.

Multiple Choice Questions

Always, *always* go through a multiple choice examination twice. The first time through, answer any questions to which you readily know the answer. Skip over the ones you don't know, or that you'll need time to figure out. Skip the ones in which you need to write out a long mnemonic to answer. That way, if you haven't completed the entire exam at the end of the allotted time, you'll have spent your time wisely in answering the ones you know. Circle the question number for the ones you don't answer. Make your circles so obvious that you can't overlook them on re-reading. Once you've completed all the questions that you know easily, go through the circled questions and attack the ones you have the best chance of answering.

Runners will often run the downhill parts of a race, and walk the uphill parts. Do the downhill parts of your examination first, and then do the slow and difficult uphill parts. Students who run out of time on an exam usually do so because they've pondered a small number of questions too long. My rule of thumb is that you should never spend more than a minute answering a question, until you've answered all the questions you can do in less than that time. Do the downhill part of your race first.

Here's the point: skipping a question on your first time through the exam isn't a bad decision if it allows you the time to answer two easier questions instead.

Sometimes on a multiple-choice question, you can eliminate two or three of the options. Cross them off on your test sheet, and then concentrate on the other two or three options. By eliminating the ones that are clearly not correct, you've eliminated distractions and increased your odds. You may have to guess, but an educated guess has a better chance of being correct than a wild guess, or no answer at all. You get the same score on the question if you answer it correctly by the process of elimination, as you get if you know the answer immediately.

Clinical Questions

We all remember "story questions" on tests in the 4th grade. Nowadays, most professional schools are emphasizing "story questions" on their examinations, because the "Board Examination" (for example the USMLE for medical students) utilizes that type of question. So you need to get conversant with that format as much as possible. Most textbooks in Anatomy include Case Studies and USMLE-type questions at the end of each chapter.

Therefore, on Anatomy examinations, you are not likely to find the question "Which of the following occurs if external laryngeal nerve is cut?" More likely you'll find the question stated this way:

A 28-year-old male presents in the Emergency Department with a neck injury due to a knife-throwing accident. The superior thyroid artery and the accompanying nerve have been severed. Which of the following can be a result of this injury?

a. loss of sensation below the level of the vocal cords

b. loss of sensation above the level of the vocal cords

c. *decreases the tension on the vocal cord on that side

d. loss of abduction function to the vocal cord on that side

e. loss of the abduction function of the vocal cords on that side

The student will need two steps in order to arrive at the answer. He or she will need to know that the nerve accompanying superior thyroid artery is the external laryngeal nerve, and they'll need to be aware that compromising external laryngeal nerve leads to decrease in tension on the vocal cord on that side.

Only time will tell us if the story question approach provides a better evaluation of students' abilities and knowledge, but it is certainly the trend in professional schools' examinations nowadays, so you'll need to be prepared for it.

I know what you're thinking: in many cases, you can skip the "story," and go right to the last sentence of the question. In poorly fabricated questions, that is indeed the case.

Occam's Razor

In answering a question on an exam, try not to say to yourself, "But what if they mean…" If you try hard enough, you can convince yourself that anything isn't so, like the people who deny the moon landing. Approach the question as simply as possible. Apply Occam's razor*. Don't assume that we professors are trying to trick you. Anatomy is a discipline in which test-makers can make difficult questions without having to resort to trickery. So read the question straight up. Be careful to avoid "reading into it" anything that isn't there.

*You will find it helpful to apply Ockham's razor to your examinations, your clinical practice, and other areas of your life, so here it is: "the simple answer is more likely to be true than one that requires several assumptions in order for it to be true." If you have to perform mental gymnastics to convince yourself that a certain answer is true, it probably isn't. *Stedman's Medical Dictionary* defines it this way: "The principle of scientific parsimony. William of Occam (14th century) stated it thus: The assumptions introduced to explain a thing must not be multiplied beyond necessity."[4] For the test-taker, don't make more assumptions than necessary. Take the question at face value.

Changing Your Answers

In grading an examination, it always upsets me to see that the student has changed his or her answer from the correct answer to an incorrect answer. In those cases, of course, I wish the student had not second-guessed his or her first answer. Is it a good idea to change an answer when you go over your exam before turning it in?

The topic of answer changing is actually an area of considerable debate. It appears that early in the literature, the debate favored students "going with their first instinct" and not changing their initial answers. However, more recent literature has actually given this a name: "the first instinct fallacy," and the pendulum has swung in the other direction, and now favors changing your answer if you are particularly uncomfortable with what you selected initially, or if you have a good reason to select another option.

Perhaps your inclination to change an answer has been influenced by some bit of information in a question farther on in the examination. Or perhaps your intuition is active, and won't stop sitting on your shoulder saying, "You know the answer you marked is incorrect. You'd better fix it."

Research suggests that when a student reviews the examination before turning it in, and changes a particular answer, it's the right thing to do. I've given this a great deal of reflection, and I suspect that, often, somewhere deep in the subconscious brain, the correct answer was lurking all the time, giving the student an uneasy feeling about the answer they selected initially.

The data are pretty clear on this. A study that I found insightful regarding students changing their answers on an examination, is published in *Teaching of Psychology*. Here are some excerpts from that article:

Staying With Initial Answers on Objective Tests: Is it a Myth?

Ludy T. Benjamin, Jr. *Texas A&M University;* Timothy A. Cavell, *Louisiana State University;* William R. Shallenberger, III, *Texas A&M University*

"Like students, most faculty (55.2%) believe that changing the initial answer will lower scores." "Most of the research in this area has been aimed at testing the accuracy of "first impressions" in test-taking. This bit of academic folk wisdom is typically stated as the belief that one should not change answers on objective tests because initial reactions to test questions are intuitively more accurate than subsequent responses."

However, the authors state that "Since 1928, at least 33 studies have been published concerning a number of issues surrounding answer- changing behavior on objective tests. Although results in these studies have sometimes been at variance, the one consistent finding is that there is nothing inherently wrong with changing initial answers on objective tests. In fact, the evidence *uniformly* indicates that: (a) the majority of answer changes are from incorrect to correct and (b) most students who change their answers improve their test scores. None of the 33 studies contradicts either of these conclusions."[5]

The point is this: if you're experiencing a nagging uneasiness with the answer that you've written, it might be an indication that the answer should be changed. Statistically, it's a good idea to change it.

Speaking of changing answers, here is an actual incident that happened at my school. At the Review Session, a student brought her fill-in-the-blank examination to my attention. The answer was correct, but had been marked wrong. Would I please correct the gradebook?

That student was surprised to learn that we make copies of all the examinations before turning them back to the students at the Review Session. It was quite obvious that she had changed her answer during the Review Session.

Therefore, as I stated above, I encourage you to change your answers from incorrect to correct, *but not after the examination has been graded.*

The Unwind

At the completion of your examination, you'll want to relax a bit. Perhaps some of your classmates will meet at a friendly venue and "debrief." You'll discuss the grossly unfair and morally corrupt exam (and examiners) over an adult beverage. We call it "The Post-Gross Toast." Here's where I'll give you another pithy bit of advice: *please party responsibly*. 'Nuff said.

After the Examination

After completing the examination and grades are published, students have the right to review the exam, to learn from their mistakes, and discover areas in which they need more work. Those opportunities are provided in a variety of formats, different at each institution. Here's how it works at my school:

Following the completion of the grading and posting process, an examination "debriefing" session is held. Course Directors are required to attend. This session provides an opportunity for the students to go over their exams. Since the examination itself is kept confidential, the review must take place under "examination conditions": the questions cannot be copied nor removed nor photographed, nor physically retained in any way. In reviewing the written examination, students are able to get clarification from the faculty on questions they missed, and in reviewing the practical examination, they can review the photo archive of all the practical examination stations. During the debriefing session, the students who still have questions or objections are invited to submit them in a written format for further review, later, by the faculty.

Cheating

It's important when taking an examination to avoid even the appearance of evil. Don't give an instructor a reason to suspect you of cheating, even if you're not cheating. As I mentioned in

Chapter ll, "How to Take Practical Examinations," don't be looking around furtively, or shielding your eyes from view, while taking written exams.

This may interest you: when I was in professional school, we had no proctors during examinations. None. Just a roomful of students (97 of us) spaced apart from each other, taking written examinations.

At the bottom of the last page of each written examination were two statements with spaces for signatures. Those statements were: "I did not cheat on this examination," and "I did not see anyone cheat on this examination." We solemnly signed our names next to each statement. We had no proctors whatsoever on those exams.

Nowadays, when I tell my students that we took examinations that way, they gasp, and then they laugh out loud. The very idea, to them, is absurd. But it is absolutely true.

When someone was accused of cheating, there was a council *comprised entirely of students*, called the Honor Council, which reviewed such cases, and recommended a course of action to the dean. That wasn't so long ago, and yet look how far we've come. I would suggest that things have not improved regarding these issues.

If you consider cheating, you should also consider this: for your professors, it's not our first rodeo; we've ridden this bull before. We were students once, too, and we've seen a classmate or two get kicked out of school for cheating, and we learned about the techniques that had been used by those cheating students. Then, when we became professors, we honed our detective skills even more. There's not much you can do regarding cheating that your professors haven't already seen.

Let's think about the student that's been caught cheating. He ran a very serious gamble (and lost) for a single point on a single examination in a professional school of some sort. He will appear before the Student Performance Review Committee (SPRC), and explain his actions to them. He may have the unhappy prospect of going home and telling his parents and his girlfriend that he's been kicked out of school for cheating. Rather than a doctor or other respected practitioner, that student could be left with an entirely different future, as a balloon bender, say, or perhaps a unicyclist.

I've heard it said that "everybody cheats nowadays, at least once in a while." It reminds me of the commuter who was driving home from work when his wife called him. "Honey, be careful. The radio says one car is going the wrong way on the freeway."

"One car?" he responded. "There's not just one car going the wrong way, but hundreds of them!"

My friend: everyone isn't cheating, and cheating is simply not worth the cost. If you decide to cheat because you believe that everyone is doing it, you may be surprised, as was the motorist going the wrong way on the freeway. Most likely you won't get away with it, and even if you do, you risked your entire professional education for want of a few points on a few exams. And your patients may suffer from your lack of knowledge. Weigh the consequences, and please don't do it. Everyone hates dealing with this: your fellow students, your professors and even the cheater. They tell me that a mouse cannot understand why the cheese in the mousetrap is free. People who think cheating is a free way to achieve their goals are as silly as that mouse.

Most graduate schools are like my school: they have a low tolerance for cheating, and that's as it should be. Modern society seems to have become comfortable with the idea that a certain number of students will cheat, and there's not much that schools can do about it. Not so in my class. If you're caught cheating in my class, the possibility exists that you'll be "dismissed with prejudice."

The Hardest Part of My Job

The most difficult aspect of my job is this: designing excellent test questions. That's a concept that never occurred to me when I was a student: all those questions that I struggled with didn't come to the instructor with ease; they were probably written several years prior, and then used every year, and were tweaked each year with changes in response to feedback from students and other faculty members. Writing sixty or more new questions per exam every year is an arduous task. That's why we don't write new questions every year; we use the good ones over and over, and that's why we develop test banks from which we select tried and true questions for quizzes and tests.

It's more difficult than you'd think to write a question with the correct level of difficulty, a question that separates the women from the girls, and the men from the boys. The question needs to be clear, but not so transparent that it's a "gimme" question. And the question must be relevant to the material that's been covered in class and/or in the assigned reading. Its content must be covered in the Objectives for the Course, and the Lecture Objectives for at least one lecture. If you think that is an easy thing to do, try to write a question that fills all the criteria above. Seriously. Go ahead and try it. I'll wait right here.

Because of those difficulties, we faculty members get together before every test, and spend an entire morning analyzing the questions on the upcoming written examination. Each question is projected on a screen, and we work together critiquing and improving each question. This process takes many hours. I've seen some instructors get defensive about their questions, and other instructors get aggressive in their criticism. Believe it or not, sometimes another professor's criticism of a question I've submitted for an examination can sting like criticism of one of my children, and any professor can be thin-skinned and tetchy in the face of incisive analysis.

You shouldn't be concerned about "tricky" questions in Anatomy. I suppose some professors intentionally write questions that are ambiguous, thinking that's what constitutes a difficult exam. However, that has been rare in my experience. Rather, my experience is that Anatomy is difficult enough that we professors can be straight-forward, writing clear questions without artifice. You should take the question as it's written, and you should not assume that the professor is trying to pull a fast one. We're not evaluating your test-taking ability; we're simply trying to ascertain whether you know the prevertebral ganglia from the paravertebral ganglia.

Practical examination stations are equally difficult to do well. At our school the students themselves set up a "Mocktical" (mock practical) examination a few days before the actual practical examination. They discover that a clear dissection, with no other structures in the way, with enough information to make clear what the question is asking, is difficult indeed. Setting up a good practical examination is as much art as science.

So you may hate examinations; you may become majorly irritable if a question is not clear, but in most cases the question wasn't written that way to intentionally confuse you. Good, clear, concise questions with the correct degree of difficulty are difficult to write. As talented as your professors are, we don't always get it right the first time.

The Dean's Philosophy

Robert Folberg, M.D., the Founding Dean at Oakland University William Beaumont School of Medicine, worked indefatigably on the concept of training compassionate physicians at his school. Many schools pay lip service to this concept; Dr. Folberg really meant it. He told his medical students, "When you study and learn at this school, we want you to de-emphasize studying for the upcoming test. Instead, we want you to focus on studying and learning *for the sake of your patients*. More important than your class tests, more important than the LCME Examination, are your patients. So let *them* be the motivation for you to learn the material comprehensively."

It's a great philosophy, no matter what healthcare field you aspire to. People smarter than you and me have determined that the stuff you learn in your professional school is important, for the sake of your patients' welfare. When you enter the cold, cruel world, or even your residency, you're going to be glad that you learned your lessons well.

Adopting Dr. Folberg's advice can give students a positive attitude in the challenges of undergraduate and professional schools. Instead of dreading that upcoming examination and its effect on his or her GPA, the student looks forward to the challenge as part of adequately preparing for real patients who will come under his or her care.

I understand that you've got admission to residency to consider. And many professional schools are pass-fail nowadays. I'm not advocating adopting this thesis exclusively, and under all circumstances. But the idea of keeping your eye on the ultimate prize: your best treatment of your prospective patients, *should be your most important motivator.*

Bradley Staats, in his excellent book *Never Stop Learning*, describes the advantage that a proper motivation can provide, as follows:

"I have a year-old schnauzer. When it was a small puppy and I wanted it to move, I kicked it in the rear and it moved. Now that I have finished its obedience training, I hold up a dog biscuit when I want the schnauzer to move. In this instance, who is motivated – I or the dog? The dog wants the biscuit, but it is I who want it to move. Again, I'm the one who is motivated, and the dog is the one who moves."

If you're sitting in the stacks at the library learning the four layers of the foot, what is your motivation for doing so? Are you more likely to be motivated to study the four layers of the foot when others tell you to learn them (as the dog does), or do you study because of your desire to be professionally competent?

Dr. Staats states it this way: "Are we more likely to be motivated when we do what others tell us, or when we behave as ourselves? In the former case, we are the dog who responds to the biscuit to follow another's motivation." He goes on to say, "When we are ourselves, the likelihood increases that we're willing to put forth the effort required to learn."[6]

In this regard, my school promotes this thought: "We study diligently, in order to improve the welfare *of our patients*." Try to keep that ultimate goal in mind. Keep your eye on the prize. Time goes by quickly, and *tomorrow*, it seems, you will happily be reaping the benefits of your effort.

What if I Fail? SPRCs and APCs

Every school has a committee that monitors student performance on an ongoing basis. The committee is called Student Performance Review Committee, or Academic Progress Committee or some similar title. When a student does not meet the standards for advancement in the program, they are required to go before the Committee and give an explanation for their disappointing performance.

You should be aware that these Committees exist *to protect the public*, and their job is to make sure that students are adequately prepared before they become healthcare practitioners. They therefore take their responsibilities seriously. At the same time, they typically approach their analysis with compassion and open minds, and a strong desire to see every student succeed.

If you're called before such a committee, you, also, should take the Committee seriously. Be prepared with a written statement, that you will read at the beginning of the session. Dress formally, displaying the fact that you realize the Committee holds your entire career in their hands. I advised one of my students about dressing with a coat and tie, and after protesting to me, he followed my advice. After the Committee hearing, he reflected, "But during the meeting, Dr. Abbot wasn't wearing a coat and tie."

To which I responded, "Dr. Abbot *passed* Gross Anatomy."

Be contrite; address Committee members by name, and above all, be prepared to explain any circumstances that compromised your ability to pass your course(s). Don't schmooze, and don't whine; *explain*. As a healthcare professional, you will frequently be attending meetings where high-stakes decisions are being made. Approach this meeting with the same professionalism that you will employ throughout your career. Preparedness, humility and gratitude should be the orders of the day.

At most schools, you will be allowed to select a faculty member to be your advocate and advisor through this process. That faculty member understands the working of the Committee, and will prepare you for the meeting. He or she already knows, or knows of, the members of the Committee, and can even inform you about what sort of attitude you can expect from each. At the meeting, your advisor does not speak, except to you, privately. You should select this counselor carefully, and then heed his or her advice assiduously.

In most cases, the protocol is this: The Committee will talk with you at length, and then you will be dismissed, and the Committee will continue to discuss the disposition of your case for as long as it takes. Ultimately, they will make a decision. Examples of possible decisions include remediation, or a curriculum modified specifically for you, or a requirement that you repeat the year, or expulsion from the school. As you leave the Committee meeting, the chairperson will tell you how and when you can expect to be notified of their decision. Traditionally, the Committee tells your advisor, and your advisor tells you.

And Finally…

I respect and admire anyone who can successfully complete a course in Gross. By the time the Midterm Exam and Final Exam come around, students in my classes have prepared hard. They've endured many hours of lecture and lab, most of the time with good humor. As a token of my appreciation for what they've accomplished, each year the last question on my Midterm Written Examination is a lagniappe:

> 100. Whew! I'm tired. Mark any answer and it will be correct:
>
> a.
> b.
> c.
> d.

Similarly, the last question on the Final Written Examination is presented as an unexpected gift:

> 100. After I finish this examination I am going to
>
> 1. hug my cadaver good-bye
> 2. punch the instructor in the nose
> 3. party responsibly
> 4. put my donor body all back together again
> 5. enjoy a restful, well-earned break

Of course, any answer is correct.

CHAPTER 12A:
Discourse with the Donor

Last night she stormed into my room, muttering something about "desiccation." Cadaver 2436, from Table 6, arrived in my bedroom at around 3:00 AM. She occasionally comes and interrupts my sleep, and talks sardonically about my negligence concerning her welfare, or some slight she's experienced in the Lab. Her demeanor is *not* the shouting, berating objurgation that commonly gives me nightmares. Instead, she usually speaks in mellifluous tones, and is interested in what's going on in my class. She's quick to point out its deficiencies (and mine).

"Yesterday, I suffered the… the absolute *deracination* of my cranium, and afterwards I was left open! *Open!* And allowed to dry out. You've got these fancy new downdraft tables, and a ventilation system that turns the air over every thirty seconds or something. That may be a whiz-bang for the Administration, but it's making me look like a shirt on a clothesline."

"The downdrafts keep the lab from smelling like formaldehyde," I said, trying to be cheerful. "The ventilation reduces the danger of that chemical to the people in the lab. Formaldehyde's a carcinogen, you know."

"You're filling *me* with carcinogens, but then those otiose students leave me, uncovered, lying there *commando*! It's ruining my entire… well, phenotype! They're supposed to turn the vents off when they leave the Lab. But they *don't!* And here's the most egregious thing!" She reached down and picked up her one remaining breast, and stretched it toward me.

I've never been more uncomfortable in my life. I momentarily stretched the sheet up over my eyes.

"My skin's dried out like a pie crust! And not just my skin, by the way," she said. She leaned over, looked deep into her pelvis, at her ovaries. At least nine of my ten pints of blood came to my face. She was right, of course, and I had no excuse.

"I've told them about keeping the cadavers moist," I said, weakly, "over and over."

"Well, when you 'tell them over and over,' it goes in one ear and out the other," she replied, "like poop through a goose."

"They're just kids," I said, my voice sounding feeble and timorous.

"It gives me wrinkles that I didn't even have in life! Dry skin! You're always preaching about how a person's wrinkles run at right angles to the pull of the muscle underneath. You never *mentioned* to the students the *cause* of the wrinkles on those of us in the lab. The windstorm in that lab makes my skin like beef jerky! You're leaving me unshrouded, and I'm drying out! Habibi, you're killing me."

The irony of that last statement was not lost on me. I suddenly sympathized with her concerns with smooth skin. Her desire was simply to be a little more gussied up. She was concerned with her appearance, as her body lies in many pieces on a cadaver table. Imagine. I suddenly had a strong desire to change the subject.

"How old, exactly, were you, when you… passed away?"

She smiled the smile of a woman keeping secrets. "Look here: the sternum end of my clavicle is fused to my manubrium. So you know I was over thirty." She chuckled to herself.

"Yes, I knew you were over thirty. Over seventy, perhaps…" She was giving nothing away. "But we can't exactly give you a massage and rub in Breck Dry Skin every evening," I said. "What are we supposed to do?"

Her tone took on a pleading quality, as she said softly and slowly, "Just close up the skin and cover me with that beach towel impregnated with formaldehyde. I'm a person who cared a great deal about my appearance when I was alive. This may surprise you," she paused before she went on, "but I used to be rather good-looking, and nowadays I look like a sculpin. I've lost my *je ne sais quoi.* The tissue around my patellae has dried out so much that my kneecaps bulge like light bulbs. I never thought that becoming a cadaver would be so… degrading."

It was quiet in my bedroom for a long time. I was sitting up, and I looked down at myself. I could stand some improvement in the looks department myself: sitting up in bed, I had rolls of abdomen piled on my lap like rice paddies on a steep hillside. She bit hard on what was left of her lower lip, and then she spoke softly. "There's a heckuva lot they didn't tell me when I signed on with this outfit." She exhaled a deep, audible breath. "Just close up the body bag, Dude. It's not rocket surgery."

"No," I responded sympathetically. "It's not rocket surgery."

She muttered quietly, "I hate that particular lab session anyway, whatever you call it, when the students remove the brain."

"I call it a lot of hard work," I said, trying to avoid the specifics.

She was chuckling as she walked toward the door. Holding the doorknob with one meticulously dissected hand, she turned and said, "One more thing. The names on your lab coats are stupid. 'Dr. Blah Blah, Ph.D.' is redundant. *Really* tacky. It should be either 'Dr.' *or* 'Ph.D.,' but definitely not both! You people have doctor's degrees and you're illiterate, already."

I told her the secretary at school ordered the lab coats with the names already embroidered. She was not persuaded. "Well, you're the professor. You should do something about it." She smiled. "Why don't you just wear your name on lanyards around your necks, like the name tags at a spelling bee? That would be more professional."

She came back to the bedside, leaned in and whispered in my ear, "Sweetheart, the way the names are written on your laboratory coats is just so… *so bush-league*." And then she vanished.

CHAPTER 13: Textbooks & Other Books

Building Your Anatomical Library

No matter what type of healthcare you plan to deliver, you'll need to know your Anatomy well. Even if you specialize in treating only the left eyes of right-handed people, your patients will expect you to know something about their TMJ dysfunction. No matter what healthcare specialty you aspire to, when you're in practice and your patient tells you that he has contracted diverticulitis, you're going to need to know what that is. You'll want to have resources at your disposal, so it's a good idea to keep a small library of your Anatomy textbooks after graduation. Make new friends, but keep the old. One is silver, and the other's *Netter*.

I've listed several books below that constitute my own canon. It's a good idea to keep some dead guys hidden away in your library. This is never more true than in your Anatomy library: you need to keep Dr. Netter, and Dr. Grant, and Dr. Gray nearby. Their wisdom never fades. If you have any intellectual curiosity at all, you'll refer to these authors throughout your lifetime. And you can acquire their books inexpensively, especially if you don't mind buying used ones. Used Anatomy atlases are affordable, even if you're living on Ramen Noodles.

There is a great number of Anatomy books out there, each with its strengths and weaknesses. If you lack discipline in shopping bookshops and Amazon, one day you'll look up from your desk, and you'll have acquired several yards of various Anatomy books on your bookshelf. That's what has happened to me. To borrow a phrase, enough Anatomy books to burn a wet mule. I hope you will become so intrigued by the subject that you will turn to other authors who can take you deeper than I ever could, into the structure and function of the human body.

The first thing you should be aware of is this: there are Anatomy *atlases*, and there are Anatomy *textbooks*, and they are not the same thing. And there is a third category, which presents anatomical case studies, like *Lachman*. I have found the last category to be helpful in learning the anatomy because these books apply the anatomy to real life situations.

If you prefer to read your books on line, most of these textbooks come in digital format, and you can readily access them in that way.

I. Essential Resources

1. Netter FH. *Atlas of Human Anatomy*, *5th ed.* Philadelphia PA: Saunders Elsevier; 2011.

Although there are many good anatomy atlases, I like Dr. Netter's. His eye is so good, his drawings so gratifyingly reliable that his atlas is comfortable equally in your library, and on your dissection table. And, in my case, my coffee table.

If you need to save money, you can buy a used copy of *Netter*, which will probably not be the most recent edition. A second-hand issue of *Netter's Atlas of Human Anatomy* is the best bargain since the Louisiana Purchase.

Johannes (or Jan) Luyken (April 16, 1649 - April 5, 1712) was a Dutch poet, illustrator and engraver. Because Mr. Luyken was an engraver as well as a poet, it would make sense that he would think that the best way to learn something is to *visualize* it. He wrote:

> "One book, printed in the Heart's own wax
> Is worth a thousand in the stacks.[1]"

When you decide to put one book in your "Heart's own wax," consider *Netter's Atlas of Human Anatomy*. However, there are many excellent candidates from which you could select your own basic atlas. Probably, yours will be the atlas required in your Gross Anatomy course. The drawings in that book will form the basis of your knowledge of the relationships and structure of human anatomy. Those drawings will be the "skeleton" on which you superimpose all other anatomical information.

Three drawbacks to *Netter*:

a. *Netter's Atlas* has no narrative at all. He just has drawings, but his drawings are among the best.

b. The publisher puts out a new edition every few years. The new editions pretty much just change the page numbers, forcing you to buy the new edition, in order to understand your professor's assignments. However, Dr. Frank Netter has passed away (a great loss to Anatomy!), and when he died he had the entire human body drawn. You can pick up a *Third Edition Netter Atlas* on Amazon for twenty bucks; it has most of the drawings that are in the later editions.

c. *Netter's* drawings can be very busy. When I assign a particular plate for my students to learn, that plate may contain many labels that aren't particularly germane to what the students and I are studying. To focus on the appropriate ones, I may assign, for a particular plate, only certain of the labeled structures in a Structures List.

2. *Stedman's Medical Dictionary, 27th ed.* Philadelphia PA: Lippincott Williams & Wilkins, 2000.

In my classes, *Stedman's Medical Dictionary* is highly recommended. I would never attempt to study Anatomy without my *Stedman's* nearby, and you can see that mine is an older edition. In addition to definitions, and etymology of the words, there are features that benefit me personally. Let me list a few:

a. Genus Finder. Have you ever seen a name of a microorganism like H. Pylori, and wondered what the "H" stands for? Check out the Genus Finder in the beginning of this book. Find "pylori," and you'll see two words, one of which starts with "H." There it is: Helicobacter. Helicobacter pylori. It's as easy as that.

b. There are three tables of Anatomy terms in the back of this book: arteries, nerves and muscles. Each reflects the *Terminologia Anatomica* nomenclature. The description of each artery tells you its origin, its course, and its branches/distribution. The muscles tell you (of course!) origin, insertion, action and innervation. The nerve section tells you the origin, course and structures innervated. These charts are amazing. They can save you a great deal of time.

c. The Building Blocks of Medical Language section lists prefixes, suffixes and combining forms for breaking down a word and figuring out what it means. This augments the eighth principle in the Forbes Method of Anatomy, called "the Forbes Method for Learning Nomenclature": *break the word down, and look it up.*

d. The Thumb Tabs help me. The tabs contain, not simply the first letter of the word at the beginning of the page; they contain the *first two* letters of that word. What a good idea!

No, I don't own stock in *Stedman's*. But I have found it to be an indispensable investment in my own Anatomy education. I know you will benefit from a good medical dictionary, and in my estimation, *Stedman's* is the best.

3. Drake RL, Vogl AW, Mitchell AWM. *Gray's Anatomy for Students, 2nd ed.* Philadelphia, PA: Elsevier; 2009.

Another book I'd buy is *Gray's Anatomy for Students*. It is a textbook, not an atlas, and therefore it has narrative descriptions as well as drawings. The narrative descriptions are basic. Those narratives can comprise the basis ("the skeleton") on which further anatomical information ("the flesh") can be hung. In contradistinction to *Netter's*, the drawings are not busy; most contain around a half-dozen labels. For the beginning Anatomy student, this book is a very valuable resource, indeed.

4. Rohen JW, Yokochi C, Lutjen-Drecoll E. *Color Atlas of Anatomy, 6th ed.* Philadelphia PA: Wolters Kluwer / Lippincott Williams & Wilkins, 2006.

For a textbook approach to actual dissections, you can't beat Rohen's *Color Atlas of Anatomy*. You can buy it new for around seventy-five bucks, and you can buy it used for as little as forty. It is a book of photographs of actual human cadaver dissections. This book will provide you pictures of your goal each time you dissect: perfect, complete, cleaned, actual dissected bodies.

5. Agur MR, Lee MJ. *Grant's Atlas of Anatomy, 10th ed.* Philadelphia PA: Wolters Kluwer / Lippincott Williams & Wilkins, 1999.

Because I am a veteran anatomist, this is an atlas that I depend on: it contains brief narrative, non-cartoonish drawings, and it is comprehensive (exhaustive!). In my opinion, this is probably not the book for a beginner, because the information is presented at an advanced level. The best thing about Grant's Atlas is the *pithy* statements at the bottom of each page. There, in short paragraphs, you will find solid information that will enlighten you. Ergo, I have great affection for this book.

6. Chung KW. *Board Review Series Gross Anatomy, 4th Edition*, Philadelphia: Lippincott Williams & Wilkins, 2000.

There are several board review books available, and this is my favorite. Reviewing questions in this book will shine a light on the areas in which you need more work.

7. Backhouse, KM. *Color Atlas of Surface Anatomy, Clinical and Applied*, Baltimore: Williams & Wilkins, 1986.

A wonderful book. It is now out of print, but when last I looked, copies are still available from Amazon and other places. If you plan any type of clinical practice, the anatomy looking from outside the body will be important. It's the basis for your physical examination of your patient. What's under the surface? Backhouse will inform you.

8. Acland, RD. *Acland's DVD Atlas of Human Anatomy*, Philadelphia PA: Lippincott Williams & Wilkins 2003.

If you have lots of money, you should buy *Acland's DVD Atlas of Human Anatomy.* It's the best movie since *Gone With the Wind.* Unfortunately, it costs around two hundred bucks for the entire series. (Used: approximately $165.00) Dr. Acland and his colleagues have done wonderful dissections of the human body. Then they have labeled and described them assiduously. This work is foundational for sure, but it provides, in video form, the basic knowledge of the relationships between structures (and Anatomy is all about relationships!)

9. Professor Marian Diamond. *Integrative Biology 131: General Human Anatomy. Fall 2005.* On You Tube.

I like to watch these videos when time will allow. You will too. Marian Diamond's lectures in Anatomy are magnificent. She has 40 lectures on You Tube, covering every structure of the body. Don't be put off by the format: blackboard (greenboard, actually) and colored chalk, simple props, and the like. These are comprehensive, engaging and wonderfully informative.

Regrettably, Dr. Diamond passed away in 2017. She was as fine an anatomy professor as there ever was, and this course is proof of that.

II. Other Books That Will Help You in Your Study:

1. Fehrenbach MJ, Herring SW. *Illustrated Anatomy of the Head & Neck*, Philadelphia PA: WB Saunders Company, 1996.

This little book is the best primer I've found for the basics of a particularly tough area of Anatomy: the head & neck. If I were starting my first exploration of the head and neck, I'd start with a quick read-through of this book. It's beautifully illustrated, and will give you the basics.

2. Kamyab A. *How to Study in Medical School, 2nd edition*, Bloomington IN: AuthorHouse , 2011

A worthwhile resource whether you're in medical school or not. This book has very little about Anatomy in particular, but I wish I had read it when I started professional school. Less than a hundred pages, and can be read in an hour, with great benefit.

3. Cahill DR. *Lachman's Case Studies in Anatomy, 4th ed.* New York NY: Oxford University Press: 1997.

When you watch ER on TV, you will hear many references to anatomy. Those references are excellent, and they are excellent in the same way that Lachman's book is excellent: here is a person who has broken his collarbone; what are the anatomical considerations, what is the treatment, what are the sequelae? *Lachman* even has a chapter on the anatomy of a hanging. Not exactly a charming topic, but you'll discover there are lots of anatomical factors that must be considered when hanging someone properly.

One of the best descriptions I've read of myocardial infarction was in *Lachman*. And real-life situations in your own experience will prompt you to pick up *Lachman* and learn the anatomy. For example, you'll encounter people who've had their gallbladder removed; *Lachman* has a chapter on it. Can a person *live* without a gall bladder? What are the consequences, and what happens to the *gall* when there is no *bladder*?

The same types of questions might be asked about a person who has had a vasectomy: when you cut the tube that takes sperm away from the testicles, do the testicles continue to manufacture sperm? If so, what happens to those sperm? Do the testicles just become bigger, and bigger, loaded down with sperm who find their exit blocked by the severed duct, until the testicle finally explodes? No, I dare say, that is not the case. But studying the anatomy of the vas deferens, and the epididymis, and the process of the vasectomy, will go a long way toward your education regarding the male reproductive system. Lachman's chapter on vasectomy begins on page 272 in my 4th edition.

4. Moore KL, Dalley AF, Agur AMR. *Clinically Oriented Anatomy, 6th ed.* Philadelphia PA: Wolters Kluwer / Lippincott Williams & Wilkins, 2010.

This will show you what a jejune life I lead: when I sit down for some enjoyable vacation reading, I like to read Anatomy. For this, I often turn to Moore, Dalley and Agur's *Clinically Oriented Anatomy.* It is the size and weight of a small water buffalo, but it is a practical textbook of Anatomy, and it's loaded with interesting clinical application.

5. Marieb EN, Mallatt J, Wilhelm PB. *Human Anatomy, 4th ed.* San Francisco CA: Pearson Benjamin Cummings, 2005.

Absolutely the best *single* book for a beginner comprehensively to learn Anatomy, in my estimation. Many times when I could not understand a complex aspect of human anatomy, I've gone to *Marieb* with good results, because *Marieb* "drills down" into so many areas that other authors do not. Her chapters dealing with the central nervous system, for example, are comprehensive on a level not found in most other common Anatomy textbooks. Also, her book is so well organized that I can usually find facts that I need without consulting the Index. Both comprehensive and comprehensible, this book is a gem. At $226.00, it's pricey.

Dr. Marieb also has a book called *Essentials of Human Anatomy & Physiology*, which adds physiology to the mix.

6. Clemente CD, *Anatomy, A Regional Atlas of the Human Body, 4th ed.* Baltimore MD: Williams & Wilkins, 1997.

Clemente has just enough narrative with each atlas drawing. His comments at the bottom of the page will help you greatly. And his drawings are elegant.

7. Hankin MH, Morse DE, Bennett-Clarke CA. *Clinical Anatomy, A Case Study Approach*, New York NY: McGraw Hill Co. Inc. 2013.

Reading Case Studies is a high-return way to study Anatomy, and Dr. Mark Hankin is a great professor. Together, those two factors make this book profitable for the Anatomy student. I encourage you to consult his book.

8. Williams PL. *Gray's Anatomy, 40th Edition*, New York, NY: Churchill Livingstone, 1995.

If you have more money than you know what to do with, and you want the most comprehensive Anatomy book, this is the book for you. It presents Anatomy on many levels: histology, embryology, pathology, on and on. Of course, *Gray's Anatomy* has been the classic since it was first published in 1858. This book is approximately the same weight as a Buick, but schlepping it around is definitely worth the physical therapy. Even my older edition, available used on Amazon for $35.07, continues to be an excellent reference book.

9. Bickley, LS, *Bates' Guide to Physical Examination and History Taking, 10th ed.* Philadelphia, PA: Wolters Kluwer / Lippincott Williams & Wilkins, 2009. Contains practical application of the anatomical knowledge you're learning.

III. Laboratory Dissectors

Many professors over the years have taken their written Laboratory Exercises and compiled them into a book appropriate for their courses. This is the provenance of most laboratory manuals. An alternative to those manuals is a professionally published laboratory manual. In my estimation, there are two professionally prepared laboratory dissectors that are fungible:

1. Tank PW. *Grant's Dissector, 14th ed.* Philadelphia PA: Wolters Kluwer / Lippincott Williams & Wilkins, 2009.

First published in 1940, now in its 14th edition, this Dissector has great drawings, dissection instructions highlighted in yellow, and important structures in boldface.

2. Clemente CD. *Clemente's Dissector, 3rd ed.* Philadelphia PA: Lippincott Williams & Wilkins, 2010.

Dr. Clemente's Dissector is stuffed with information. It is a dissector for the serious advanced student, and its descriptions are most suitable for a professional school. A comprehensive dissector.

IV. Reading Anatomy for Fun

If you love Anatomy, and/or if you are curious about it, try some of these for your summer reading next year:

1. McNeill D. *The Face, A Natural History.* Boston MA: Little, Brown & Company, 2000. ISBN 0-316-58812-1.

This is a wonderful little book. Mine is all marked up and dog-eared because I've read it several times.

2. Swenson RA. *More Than Meets the Eye*, Colorado Springs, CO: Navpress, 2000.

Are you amazed by the human body? You ain't seen nuthin' yet. Take this book on vacation, and you'll find yourself reading portions aloud to people around you, even on the subway. And you'll get a sense of the divine influence on the formation of your body.

3. Cahill DR. *Lachman's Case Studies in Anatomy, 4th ed.* New York NY: Oxford University Press: 1997.

You may notice that I have this book listed under "2. Other Books That Will Help You in Your Study," above, as well as here at "4. Reading Anatomy for Fun." If you're interested in clinical application of Anatomy, this book will educate you and entertain you. It presents cases, and then explains the outcomes comprehensively.

4. Nuland, SB. *The Mystery Within, A Surgeon Explores Myth, Medicine, and the Human Body*, New York, NY: Simon and Schuster: 2000.

My favorite author on medical practice.

5. Hayes B *The Anatomist, A True Story of Gray's Anatomy*, New York, NY: Bellevue Literary Press, 2009.

This is a great book, about the provenance of a great book.

V. Books by Laymen About the Cadaver Lab Experience

1. Giegerich S. *Body of Knowledge, One Semester of Gross Anatomy, The Gateway to Becoming a Doctor*, New York, NY: Simon and Schuster, 2001.
2. Carter III AH. *First Cut, A Season in the Human Anatomy Lab*, New York, NY: Picador USA, 1997.
3. Montross C. *Body of Work, Meditations on Mortality From The Anatomy Lab*, New York, NY: Penguin Books, 2007.

4. Roach, M. *Stiff, The Curious Lives of Human Cadavers*, New York, NY: W.W. Norton, 2003.

VI. Anatomy Books for the Layman

Lots of anatomical knowledge, written in language comfortable for the layman:

1. Taylor JB. *My Stroke of Insight, a Brain Scientist's Personal Journey*, New York NY: The Penguin Group, 2006.
2. Swenson RA. *More Than Meets the Eye, Fascinating Glimpses of God's Power and Design.* Colorado Springs CO: Navpress, 2000.
3. Gawande A. *Complications A Surgeon's Notes on an Imperfect Science*, New York, NY: Picador, 2002.
4. Baden M. *Dead Reckoning, The New Science of Catching Killers*, New York, NY, Simon and Schuster, 2001.
5. Nuland SB. *The Mysteries Within, A Surgeon Explores Myth, Medicine, and the Human Body*, New York NY: Simon and Schuster, 2000.

Sherwin B. Nuland is one of my favorite authors. He has another book named *The Wisdom of the Body.* Wonderful reads, for the anatomically curious.

VII. Layman Books That Will Add to Your Anatomy Knowledge

1. Verghese, Abraham. *Cutting for Stone*, New York NY: Vintage Books, 2009.
2. Hayes B. *The Anatomist, A True Story of Gray's Anatomy*, New York, NY: Bellevue Literary Press, 2009.
3. Siebert, C. *A Man After His Own Heart, A True Story*, New York, NY: Crown Publishers, 2004.

VIII. Books about Learning and Memorizing in General

1. Kamyab A. *How to Study in Medical School, 2nd ed.* Bloomington IN: AuthorHouse, 2011.
2. Lorayne H, Lucas J. *The Memory Book*, New York NY: Barnes & Noble Books, 1974.

3. Foer J, *Moonwalking with Einstein, The Art and Science of Remembering Everything*, New York, NY: The Penguin Press, 2011.

4. Newport, C. *How To Win At College, Surprising Secrets for Success From The Country's Top Students*, New York, NY: Broadway Books, 2005.

5. Staats, BR, *Never Stop Learning, Stay Relevant, Reinvent Yourself, and Thrive*, Boston, MA: Harvard Business Review Press, 2018.

IX. Other Valuable Resources

DVD: (Acland) Acland RD. *Acland's DVD Atlas of Human Anatomy*, Philadelphia PA: Lippincott Williams & Wilkins 2003.

You Tube: *Integrative Biology 131: General Human Anatomy*. Fall 2005. Professor Marian Diamond.

Marian Diamond's lectures in Anatomy are magnificent. She has 40 lectures on You Tube, covering every structure of the body. Don't be put off by the protocols: blackboard and colored chalk, simple props, and the like. These are comprehensive, engaging and wonderfully informative. You'll like them.

X. An Exhaustive List of My Favorite Anatomy Textbooks

Backhouse KM, Hutchings RT. *Color Atlas of Surface Anatomy, Clinical and Applied.* Baltimore MD: Williams & Wilkins; 1986.

Bickley LS, Szilagyi PG. *Bates' Guide to Physical Examination and History Taking, 10th ed.* Philadelphia PA: Wolters Kluwer / Lippincott Williams & Wilkins, 2009.

Clemente CD. *Anatomy, A Regional Atlas of the Human Body, 4th ed.* Baltimore MD: Williams & Wilkins; 1997.

Clemente CD. *Clemente's Dissector, 3rd ed.* Philadelphia PA: Lippincott Williams & Wilkins, 2010.

DeBrul EL. *Sicher's Oral Anatomy, 7th ed.* St. Louis MO: The C.V. Mosby Company, 1980.

Evers H, Haegerstam G. *Introduction to Dental Local Anesthesia*, Fribourg Switzerland: Mediglobe SA, 1990. The drawings in this book are elegant.

Moore KL, Agur AMR. *Essential Clinical Anatomy, 3rd ed.* Philadelphia PA: Lippincott Williams & Wilkins, 2007.

Fehrenbach MJ, Herring SW. *Illustrated Anatomy of the Head and Neck, 2nd ed.* Philadelphia, PA: W.B. Saunders Company, 2002.

Tank PW. *Grant's Dissector, 14th ed.* Philadelphia PA: Wolters Kluwer / Lippincott Williams & Wilkins, 2009.

Agur AMR, Lee MJ. *Grant's Atlas of Anatomy, 10th ed.* Philadelphia PA: Lippincott Williams & Wilkins; 1999.

Drake RL, Vogl AW, Mitchell AWM. *Gray's Anatomy for Students, 2nd ed.* Philadelphia, PA: Elsevier; 2009.

Cahill DR. *Lachman's Case Studies in Anatomy, 4th ed.* New York NY: Oxford University Press: 1997.

Lang J. *Clinical Anatomy of the Masticatory Apparatus and Peripharyngeal Spaces.* New York NY: Thieme Medical Publishers; 1995.

Larrabee WF, Makielski KH, Henderson JL. *Surgical Anatomy of the Face, 2nd ed.* Philadelphia PA: Lippincott Williams & Wilkins; 2004.

Marieb EN, Mallatt J, Wilhelm PB. *Human Anatomy, 4th ed.* San Francisco CA: Pearson Benjamin Cummings, 2005.

McMinn RMH, Hutchings RT. *Color Atlas of Human Anatomy, 2nd ed.* Chicago, IL: Yearbook Medical Publishers, 1988.

Logan BM, Reynolds PA, Hutchings RT. *McMinn's Color Atlas of Head and Neck Anatomy, 3rd ed.* Edinburgh, UK: Yearbook Medical Publishers, 2004.

Moore KL, Dalley AF, Agur AMR. *Clinically Oriented Anatomy, 6th ed.* Philadelphia PA: Wolters Kluwer / Lippincott Williams & Wilkins, 2010.

Ide Y, Nakazawa K, Hongo T, Tateishi J. *Anatomical Atlas of the Temporomandibular Joint.* Tokyo, Japan: Quintessence; 2001. A real eye-opener.

Netter FH. *Atlas of Human Anatomy, 5th ed.* Philadelphia PA: Saunders Elsevier; 2011.

New Oxford American Dictionary, 3rd ed. New York, NY: Oxford University Press: 2010.

Paff GH. *Anatomy of the Head and Neck.* Philadelphia PA: WB Saunders Co., 1973. Out of print, but worth the effort if you can find one.

Rohen JW, Yokochi C, Lutjen-Drecoll E. *Color Atlas of Anatomy, 6th ed.* Philadelphia PA: Wolters Kluwer / Lippincott Williams & Wilkins, 2006.

Snell RS. *Clinical Anatomy by Regions, 8th ed.* Philadelphia PA: Wolters Kluwer / Lippincott Williams & Wilkins 2008.

Stedman's Medical Dictionary, 27th ed. Philadelphia PA: Lippincott Williams & Wilkins, 2000.

Swenson RA. *More Than Meets the Eye*, Colorado Springs CO: Navpress, 2000.

(Wb) Woodburne RT, Burkel WE. *Essentials of Human Anatomy, 9th ed.* New York NY: Oxford University Press; 1994.

CHAPTER 13A: Discourse with the Donor

"Felicitations!" she said. "The end of the semester is here."

She arrived with a cheerful demeanor at my bedside as she was wont to do, but we both knew this visit would be her last. Her remaining skin was the color of milk. She sounded far away, and the timbre of her voice was like the sound of tires on a gravel road. "Time to turn in my badge and gun," she said. She looked down at her perfectly dissected abdomen. "I'm completely spent," she said quietly.

And she was. The team of students at her dissection table had done a masterful job throughout the semester, and many parts of her anatomy resembled the dissections in *Rohen*. No one was surprised when the group at her table won the coveted "I Had the Best Body" T-shirts, for excellence in dissection. Throughout the semester, this donor had occasionally appeared at my bedside, with surprising responses to what was going on in the lab. And she had helped me address the flotsam and jetsam of my own myopia regarding the proper administration of a cadaver lab. And she was not loath to give me her two cents regarding my relationships with my beloved students.

And now, I knew in my heart that this was her last such visit. I missed her already, and she wasn't even gone yet.

"Cheer up!" she said, "You look like a Pernkopf dissection. You knew this day was coming. It comes with the territory." She glided about the room, like a balloon in a gentle wind. "By the way, on the Final Exam yesterday, you didn't allow enough time at each station. The students were majorly pressed for time. I thought you told me you didn't write tricky questions on examinations."

"Tricky questions!" I was offended. Being accused of dodginess with my test questions is not something I take lightly. That's a real insult to an Anatomy professor. "There were no tricky questions. The questions on my examinations simply require the student to come up with the answer without folderol. I'm testing their knowledge of Anatomy."

"When students don't have enough time to examine the cadaver and reflect on the question, you're not testing their knowledge of Anatomy. You're testing their reflexes, their mental celerity. That's what I call a tricky question."

"We give them five rest stations in the course of the examination, and a lightning round at the end, with twenty seconds per station. That gives them plenty of time to reflect."

"I'm just telling you what I observed. You've told me yourself that you don't do your best work when you hurry, and it just might be the same for the students."

Once again, she was right. I resolved to allow more time in the future. "I will miss you," I said weakly. This wasn't the first time I'd learned something important from her. She had developed proficiency in Anatomy, as well; in some cases I had found her even more helpful than the written Anatomy atlas. Perhaps it was because *this* atlas had a pulse. *Or did she*?

As I was pondering the fact that not long ago she *did* have a pulse, and a blood pressure, my cadaver friend displayed some pretty serious hypertension about another issue. Her neighbor's anterior triangle had been dissected poorly, and then had been used as a test station on the examination. Another professor had done the dissection, and in the process he had severed the ansa cervicalis nerve that drapes the front of the carotid sheath like a gold chain around someone's neck. He had attempted to approximate the two severed ends, but the repair was clunky and obscure. I already knew about the problem; on their way out, two students had complained to me about its poor quality.

The poor examination station had upset me, as well. The Final Examination in Gross Anatomy is a high-stakes examination in medical school, and a missed question one way or the other can make a big difference to a student. Earlier, I had laid awake in bed reliving, in Sisyphean redux, my failure in supervision. "Dr. Midias made a mistake," I admitted. "I asked him to reflect the sternocleidomastoid muscle, and I neglected warning him about the ansa on the sheath. I must be getting too old for this."

"You can't be everywhere at once," she consoled me. "He should have known." Then she smiled with the undissected half of her mouth, an asymmetric smile that turned into a leer. "I liked what you said as the students were leaving the lab after the final examination."

"What was that?"

"You acknowledged that they deserve a celebration the end of the course, and you said, 'Please party responsibly!' I like that. You really do like the students, don't you?"

"Of course I do; I love the students," I responded. "They have just spent upwards of five hundred hours studying and learning a subject that is near and dear to my heart. Along the way, each one has earned my respect. What's not to love? So obviously I don't want them to celebrate excessively at the completion of the course, and possibly end up dead."

Silence.

"No," she said quietly. "We wouldn't want that."

I was completely gobsmacked by my own rudeness. I rushed to repair it like snatching up a toddler in traffic. "I'm sorry if I offended you. I was referring to the students, many of whom are in their early 20s, and…"

"No offense taken, Doctor," she said.

I couldn't let it go. "I'm simply mortified about what I said."

Looking at her, my eyes were suddenly like looking through greasy windows. When she spoke, it was beginning to sound like listening to someone in the next room, using a drinking glass pushed up against the wall. "If you think about it, Doctor, I'm the one who's 'mortified,'" she said, smiling. "I'm afraid my body long ago lost that inevitable last argument with nature. It's time to give in." She looked me in the eyes. "I fought the good fight, I've finished the course. I have kept the faith."

A surge, like a massive overdose of Ethiopian coffee, passed through my body. My heart fluttered in my chest like a moth in a jar. She was leaving. She headed toward the door. Then suddenly she stopped, turned back and said, "Don't ever forget me, Bill." And then my friend was gone.

CHAPTER 14:
The Memorial Service (The Service of Appreciation)

A Rich Variety of Farewells

Most schools with a cadaver laboratory have a Memorial Service honoring the donors at the end of the Anatomy course. These services take various forms. In some, families of the donors are invited to attend, students wear white coats, and faculty dress appropriately. Schools will seek a commodious venue: an auditorium at the school, or a church nearby. Generally, there is a reverent atmosphere. Often a student has written a poem about, or addressed to, the donor. Music is provided by students, vocal and/or instrumental. A representative of the school brings some remarks. The year I was invited to speak, it was a great honor.

At one school, there is a short ceremony in the anatomy lab itself, where a rose is placed on each body encased in a spotless body bag, ready to return to the morgue. Each student is allowed the opportunity to make a statement, read a Scripture reference, or read a poem. Students are invited to say "Thank you" in any language they can speak. I've had as many as twenty-five different languages spoken by students in some of my classes.

Each school is creative in presenting a program honoring the donors. The program often focuses on importance of the gift to the students' education and to the patients they will treat throughout their lifetimes.

The notes for my address to the assembled multitude are written below.

Good evening, my friends.

It is a remarkably generous thing to do, to donate your body to science.

And it's a blessing that there are those among us who leave us this amazing gift, to study in the Anatomy Lab.

We're here this evening to honor those who donated their bodies. Without their courage and generosity, we wouldn't have the opportunity to study an actual human body, and there is simply no substitute for that.

I'm told that one motivation to make this gift is in service to humanity, and I believe we are justified in doing so.

Let's see: there are typically five or six students at each laboratory table, and each of you will go on to use his or her knowledge

to treat thousands of patients in their lifetime, each patient in turn with a family hoping and praying for a good outcome from the treatment.

Therefore, it's not unreasonable to conclude that each donor body will, indirectly, improve the lives of thousands of people.

So we honor their decision, and we treat those donated bodies with care and respect.

And there's another reason we treat the bodies this way. There is, without a doubt, a divine aspect to the study of the human body. The Bible says that we are "fearfully and wonderfully made," that we are created in God's own likeness. It goes on to say that our bodies are nothing less than the temple of the Holy Spirit of God.

If anyone has worked with these human bodies and escaped the notion that there is something sacred in that work, they have missed a great deal, in my view.

People lived in these bodies, assiduously caring for them, their whole lives.

And in the future, as healthcare providers, you will participate in caring for your living patients – advising them, helping them, treating them.

In our work in the Anatomy Laboratory, we act out a major principle of this school, and it's this: the reason we study is in order to better serve our patients. The reason we study in the Anatomy laboratory is in order to <u>*better serve our patients.*</u>

My friends, if better serving our patients in the care of their bodies is not a divine pursuit, I don't know what is.

The presence of our donor bodies reminds us of the fleeting nature of life, and being so reminded, it prompts us to be kind, and considerate of one another. No matter what stage of life, our time is short, which makes us more circumspect, more thoughtful, in our discourse.

And what an astonishing thing it is, to study the human body! How many times during the course of our time together in the laboratory did we say, "Wow!," or "Holy Cow!"

The ovary throws out her ovum into the vastness of the peritoneum, to be snagged by a device resembling a baseball glove. There, the ovum enters a long tube, and travels down, down, down. Before long she meets a nice young man coming from the opposite direction. They get married in that tube, and then they travel, holding hands, into the uterus to begin the happy event of child-rearing.

Could anything be more amazing? Could anything be more… ***bodacious****? Over and over, I come away from my study of human anatomy amazed by miracles that happen all the time, in these bodies we're blessed to live in; don't you?*

Dr. Folberg, the Dean, has spent his professional life focused on the eye, which, as we all know all too well, is a structure of amazing sensitivity and complexity.

Think for a moment about entering a dark movie theater, and how your eye, without any conscious effort from you, accommodates to the low light.

Then, you look up and your eyes focus perfectly on the screen. Then you look down, and your eyes focus clearly on your popcorn. Your eyes just go about doing thousands of processes, thousands of adjustments, completely independent of your conscious help.

And how about this: tell me another structure that compares with this: in eighty years of life, our eyelids will blink, flawlessly, 400 million times.

What can compare with it? It's amazing beyond belief.

The human heart is made of muscle. The muscle contracts and beats one hundred thousand times each day sometimes for an entire century. When does that muscle take a rest?

Sometimes I envision the time when my remains are lying on the dissection table, and the group of students will open up my chest and peer in.

I believe that when they do, they will suddenly stop, and gasp, and then lean in for a closer look. There, on the surface of the ventricles, they will see a startling sight. They will see that many of your names are plainly written there.

I've loved working in Gross Anatomy since before you were born, and I've loved working with people like you all, doing work so meaningful, so interesting, so… well… engrossing that I could hardly get enough of it.

Gross Anatomy is the floor of many rooms, and those rooms are embryology, and physiology, and pathology, and histology, and biochemistry, and on and on.

We learn the foundations of the human body in the cadaver lab. And when we dissect, we're teaching Anatomy to our hands.

We begin that study as strangers – strangers to the human body and strangers to each other - and we invariably finish as good friends to both.

I'm so glad you haven't missed this rare opportunity.

And someday, when you're really old, (like me) twenty, thirty, forty years after you graduate, when you've achieved your goals and you are a practicing healthcare professional, you will get a little whiff of formaldehyde, and you will remember the Anatomy Lab, the names of your lab partners, and your instructors, and your donor body. And you'll remember what I call the "music" of the Anatomy Lab: the sound of learning, the sound of discovery, the sound of students teaching other students all throughout the room. And you will recall that lovely aroma in the lab.

And here's the thing: when you remember, you will smile. I promise you will.

For most people, what's inside the human body is a great mystery. Haven't you found that to be true? They entertain vague notions about where organs are located, what they do and how they work.

But for those of us who have the splendid opportunity to explore it for ourselves, to actually see for ourselves the beauty and grace of the human body, and to learn its secrets in order to better serve our patients, it's a compelling, once-in-a-lifetime experience.

Each of us is unique, that's obvious, but when we get inside, we find that we're all quite the same. Ours is a uniquely privileged study, and that study would be impossible except for the unrequitable thoughtfulness of people we've never met,

who had the charity to give to us what was their most intimate home for seventy, eighty, ninety years.

My dear friends: the only appropriate response is profound gratitude. Thank you.

ENVOI:
The Afterword

Serving the University as a Gross Anatomy professor has always made me proud. Teachers and students are partners in a dance of great importance. One of the great rewards for superannuated teachers is finding the occasional opportunity to join the dance once again. Writing this book, *Learning Gross*, has been that, for me. As Parker J. Palmer wrote in his thoughtful book *On the Brink of Everything*: "It is the dance of the spiraling generations, in which the old empower the young with their experience and the young empower the old with new life, reweaving the fabric of the human community as they touch and turn."[1] Students in my classes have done a great service to this professor, with their courtesy, their apt attention, and their enthusiasm for a discipline in which I have been immersed my entire life.

Nowadays there's a happy experience when I happen to meet a former student: no matter how many years intervene between our class together and our chance meeting in the future, when you introduce me, you say, "Dr. Forbes was my teacher." It is the highest metier I've ever aspired to, and it's the fulfilment of all my intentions: to share with you some fragments of understanding, as together you and I regard, with astonishment, the gross anatomy of the human body.

REFERENCES

Preface

1. Brann, E. (2014). *Un-willing: An inquiry into the rise of will's power and an attempt to undo it.* Paul Dry Books.
2. Oxford University Press. (2010). *The New Oxford American Dictionary (3rd ed.*, p. 1170*)*

Chapter 1: Introduction to Gross Anatomy

1. Giegerich, S. (2001). *Body of knowledge: One semester of gross anatomy, the gateway to becoming a doctor.* (Title page). Scribner.

Chapter 2: Relationships with Professors

1. Oxford University Press. (2010). *The New Oxford American Dictionary (3rd ed.*, p. 1564*)*
2. Hayes, B. (2009). In *The anatomist: A true story of gray's anatomy* (p. 85). Bellevue Literary Press.
3. Taitz, J. L. (2018, August 16). The Case Against Cutting Class. *Wall Street Journal*, p. A13.
4. Augustine, Pusey, E. B., & Symons, A. (1906). *The confessions of saint Augustine.* Feeling-on-Tyne. Retrieved April 19, 2023 from **https://www.ccel.org/ccel/augustine/confess.xi.viii.html**

Chapter 3: Anatomical Terminology, Etymology and Abbreviations

1. Oxford University Press. (2010). *The New Oxford American Dictionary (3rd ed.*, p. 249*)*

2. Stedman, T. L. (2000). *Stedman's medical dictionary (27th ed.*, p. 1482*)*. Lippincott Williams & Wilkins.
3. Oxford University Press. (2010). *The New Oxford American Dictionary (3rd ed.*, p. 753*)*
4. Stedman, T. L. (2000). *Stedman's medical dictionary (27th ed.*, p. 1797*)*. Lippincott Williams & Wilkins.
5. Bernstein, T. M. (1965). *The careful writer: A modern guide to English usage* (p.96). Free Press/Simon and Schuster.
6. Stedman, T. L. (2000). *Stedman's medical dictionary (27th ed.*, p. 1429*)*. Lippincott Williams & Wilkins.
7. Stedman, T. L. (2000). *Stedman's medical dictionary (27th ed.*, p. 611*)*. Lippincott Williams & Wilkins.
8. Stedman, T. L. (2000). *Stedman's medical dictionary (27th ed.*, p. 1082*)*. Lippincott Williams & Wilkins.
9. Oxford University Press. (2010). *The New Oxford American Dictionary (3rd ed.*, p. 70*)*
10. Stedman, T. L. (2000). *Stedman's medical dictionary (27th ed.*, p. 1643*)*. Lippincott Williams & Wilkins.
11. Stedman, T. L. (2000). *Stedman's medical dictionary (27th ed.*, p. 579*)*. Lippincott Williams & Wilkins.
12. Stedman, T. L. (2000). *Stedman's medical dictionary (27th ed.*, p. 1059*)*. Lippincott Williams & Wilkins.
13. Oxford University Press. (2010). *The New Oxford American Dictionary (3rd ed.*, p. 1360*)*
14. Stedman, T. L. (2000). *Stedman's medical dictionary (27th ed.*, p. 1295*)*. Lippincott Williams & Wilkins.
15. Stedman, T. L. (2000). *Stedman's medical dictionary (27th ed.*, p. 332*)*. Lippincott Williams & Wilkins.

Chapter 3a. Discourse With The Donor

1. Shakespeare, W. *The Merchant of Venice*, Act-IV, Scene-1. Retrieved on April 19, 2023, from **https://poets.org/poem/merchant-venice-act-iv-scene-i-quality-mercy-not-strained**

Chapter 4: Acquiring Knowledge

1. Swenson, R. A. (2000). *More than meets the eye: Fascinating glimpses of god's power and design.* (p. 93) NavPress.
2. Hayes, B. (2009). *The anatomist: A true story of gray's anatomy.* (p. 8) Bellevue Literary Press.
3. Foer, J. (2011). *Moonwalking with einstein: The art and science of remembering everything.* (p. 44) Penguin Books.
4. Stedman, T. L. (2000). *Stedman's medical dictionary (27th ed.*, p. 1590*)*. Lippincott Williams & Wilkins.
5. Foer, J. (2011). *Moonwalking with einstein: The art and science of remembering everything.* (p. 62) Penguin Books.
6. Taylor, J. B. (2009). *My stroke of insight: A brain scientist's personal journey.* (p.16) Used by permission of Viking Books, an imprint of Penguin Publishing Group, a division of Penguin Random House LLC. Plume.
7. Oxford University Press. (2010). *The New Oxford American Dictionary (3rd ed.*, p. 1122*)*
8. Scalia, A., Scalia, C. J., & Whelan, E. (2017). *Scalia speaks: Reflections on law, faith, and life well lived.* (p. 329) Crown Forum.
9. Kamyab, A. (2011). *How to study in Medical School (2nd ed.)*. AuthorHouse.

Chapter 5: How to Study Gross Anatomy

1. Staats, B. R. (2018). *Never stop learning: Stay relevant, reinvent yourself, and thrive.* (p. 140) Reprinted with permission from Harvard Business Review Press.

2. Carey, B. (2010, September 6). *Forget what you know about good study habits.* The New York Times. Retrieved April 17, 2023, from https://www.nytimes.com/2010/09/07/health/views/07mind.html

3. Stedman, T. L. (2000). *Stedman's medical dictionary (27th ed.*, p. 72*)*. Lippincott Williams & Wilkins.

4. Backhouse, K. M., & Hutchings, R. T. (1992). *Colour atlas of surface anatomy, clinical and applied.* Wolfe Medical.

5. Fehrenbach, M. J., & Herring, S. W. (2020). *Illustrated anatomy of the head and Neck.* Elsevier/Saunders.

6. Lillie, J. H., & Bauer, B. A. (1994). *Sectional anatomy of the head and neck: A detailed Atlas.* Oxford University Press.

Chapter 6: The Forbes Method of Anatomy

1. Staats, B. R. (2018). *Never stop learning: Stay relevant, reinvent yourself, and thrive.* (p. 42) Reprinted with permission from Harvard Business Review Press.

2. Kamyab, A. (2011). *How to study in Medical School* (2nd ed.). AuthorHouse.

3. Kamyab, A. (2011). *How to study in Medical School* (2nd ed. p. 13). AuthorHouse.

4. Newport, C. (2005). *How to win at college: Surprising secrets for success from the country's top students.* (p. 175) Three Rivers Press.

5. Staats, B. R. (2018). *Never stop learning: Stay relevant, reinvent yourself, and thrive.* (p. 154) Reprinted with permission from Harvard Business Review Press.

6. Staats, B. R. (2018). *Never stop learning: Stay relevant, reinvent yourself, and thrive.* (p. 166) Reprinted with permission from Harvard Business Review Press.

7. Rohen, J. W., Yokochi, C., Lutjen-Drecoll Elke, & Romrell, L. J. (2006). *Color atlas of anatomy a photographic study of the human body* (6th ed.). Schattauer.

8. Agur, A. M. R., & Lee, M. J. (1999). *Grant's Atlas of Anatomy* (10th ed.) Wolters Kluwer.

9. Drake, R. (2019). *Gray's Anatomy for Students.* (4th ed.) Elsevier.
10. Netter, F. (2011). *Atlas of Human Anatomy.* (5th ed.) Saunders/Elsevier
11. Clemente, C. D. (2012). *Anatomy: A regional atlas of the human body* (6th ed.) LWW.
12. YouTube. (2007). *Integrative Biology 131: General Human Anatomy. YouTube.* Retrieved April 17, 2023, from https://www.youtube.com/watch?v=XGS-gGsyMTw.
13. Oxford University Press. (2010). *The New Oxford American Dictionary (3rd ed.*, p. 1520*)*
14. Oxford University Press. (2010). *The New Oxford American Dictionary (3rd ed.*, p. 1345*)*
15. Swenson, R. A. (2000). *More than meets the eye: Fascinating glimpses of god's power and design.* (p. 36) NavPress.

Chapter 7: The Cadaver Laboratory

1. New York Times (1882, December 20). The Deeds of Criminals: Dr. Forbes Indicted—Greenwald Confesses—Sentenced for Life. (1882, December 20). *New York Times.*
2. Hayes, B. (2009). *The anatomist: A true story of gray's anatomy.* (p. 166) Bellevue Literary Press.
3. Hayes, B. (2009). *The anatomist: A true story of gray's anatomy.* (p. 167) Bellevue Literary Press.
4. Hayes, B. (2009). *The anatomist: A true story of gray's anatomy.* (p. 30) Bellevue Literary Press.
5. Marieb, E. N., Mitchell, S. J., & Zao, P. Z. (2016). *Human Anatomy & Physiology Laboratory Manual: Cat Version.* Pearson.
6. Austen, J. (1996). *Pride and prejudice.* Penguin Books. First published 1813.

Chapter 8: The Cadaver Laboratory Protocols

1. Stedman, T. L. (2000). *Stedman's medical dictionary (27th ed.*, p. 497*)*. Lippincott Williams & Wilkins.

2. Dalrymple, T. (2018, December) *Letter to an aspiring doctor.* First Things. Retrieved April 19, 2023, from https://www.firstthings.com/article/2018/12/letter-to-an-aspiring-doctor#:~:text=You'll%20make%20mistakes%2C%20but,but%20the%20harm%20often%20hidden.

3. Staats, B. R. (2018). *Never stop learning: Stay relevant, reinvent yourself, and thrive.* (p. 159) Reprinted with permission from Harvard Business Review Press.

4. Shakespeare, W. *Othello*, Act 5, Scene 2. Retrieved April 29, 2023, from https://nosweatshakespeare.com/quotes/soliloquies/it-is-the-cause/

5. Oxford University Press. (2010). *The New Oxford American Dictionary (3rd ed.*, p. 1908*)*

6. Shockley, W. Retrieved April 19, 2023 from https://www.azquotes.com/quote/1310399

Chapter 9: The Cadaver

1. Shakespeare, W. *Hamlet* 3.1.56-69. Retrieved on April 19, 2023 from https://www.shmoop.com/shakespeare-quotes/mortal-coil/

2. Stedman, T. L. (2000). *Stedman's medical dictionary (27th ed.*, p. 1574*)*. Lippincott Williams & Wilkins.

3. Stedman, T. L. (2000). *Stedman's medical dictionary (27th ed.*, p. 1025*)*. Lippincott Williams & Wilkins.

Chapter 10: How to Dissect

1. Oxford University Press. (2010). *The New Oxford American Dictionary (3rd ed.*, p. 611*)*

2. Hayes, B. (2009). *The anatomist: A true story of gray's anatomy.* (p. 85) Bellevue Literary Press.

3. Vincent, F. (2019, July 19). A Good Life in My Dying Days. *The Wall Street Journal*, p. A15.

4. Montross, C. (2008). *Body of work: Meditations on mortality from the Human Anatomy Lab.* Penguin Books.

Chapter 11: How to Take Practical Examinations

1. Marieb, E. N., Brady, P., & Mallatt, J. (2005). *Human anatomy.* Pearson Education Limited.

2. Ziskin, L., & Schulman, T. (1991). *What about Bob?* United States; Buena Vista Pictures Distribution, Inc.

Chapter 12: How to Take Written Tests

1. Ackerman, J. (2007). *Sex sleep eat drink dream: A day in the life of your body.* (p. 184) Houghton Mifflin.

2. Chung, K. W. (2005). *Gross anatomy* (5th ed.). Lippincott Williams & Wilkins.

3. Staats, B. R. (2018). *Never stop learning: Stay relevant, reinvent yourself, and thrive.* (p. 108) Reprinted with permission from Harvard Business Review Press.

4. Stedman, T. L. (2000). *Stedman's medical dictionary (27th ed.*, p. 1250*)*. Lippincott Williams & Wilkins.

5. Benjamin, L. T., Cavell, T. A., & Shallenberger, W. R. (1984). Staying with initial answers on objective tests: Is it a myth? *Teaching of Psychology, 11(3)*, 133–141. **https://journals.sagepub.com/doi/10.1177/009862838401100303**

6. Staats, B. R. (2018). *Never stop learning: Stay relevant, reinvent yourself, and thrive.* (p. 104) Reprinted with permission from Harvard Business Review Press.

Chapter 13: Textbooks & Other Books

1. Foer, J. (2011). *Moonwalking with einstein: The art and science of remembering everything.* (p. 110) Penguin Books.

Envoi: The Afterword

1. Palmer, P. J. (2019). *On the brink of everything: Grace, gravity, and getting old.* (p.38) ReadHowYouWant.

ACKNOWLEDGEMENTS

Many thanks to all the members of my immediate family, who helped me immeasurably on this project. I'm grateful for your encouragement - no, your insistence - that I finish this book and get it published: Gerri Forbes, Karl and Katherine Forbes-Smith, Sarah and Kevin Grant, Heather and Bill Forbes. Each of you contributed in different ways – all significant ones – in this arduous process, and your influence is laced throughout this book. Early on, Bill performed a ruthless and consummate proofreading of the entire text, improving the book tremendously. I owe a huge debt of gratitude to everyone in this extraordinary group, who are, thankfully, also my treasured friends. God has once again proven that He is able to do exceedingly abundantly above all that I could ever ask or think (Ephesians 3:20).

INDEX

G

H

I

J

K

L

M

N

O

P

Q

R

S

T

APPENDIX I: The Anatomy Course Documents

Most Anatomy course schedules are divided into segments. Commonly there are four: (1.) upper extremity, (2.) torso, (3.) lower extremity & pelvis, and (4.) head & neck. Your course may be different. Especially if yours is a dissection course, there are certain sequences that simply make more sense. In my courses, we'd like to begin by dissecting in an area that has big structures and simple patterns of anatomy. So the back, or the lower extremity are natural places to start.

I have included below the Syllabus for the Head & Neck portion of the Gross Anatomy course I taught for many years. There are things to which I call your attention:

There is a quiz to begin each lecture, covering basics of the material that *will be* discussed in that lecture. The students are expected to review assigned pages, come to class and take an (easy) quiz, in order to prove that they have looked over the material prior to our discussion. Forcing the students to acquire a basic understanding of the material *before* the lecture allows me to use my scarce and valuable lecture time this way: skip the basics ("This is the thyroid cartilage; here is the external laryngeal nerve, that's cricothyroid muscle."), and instead talk about the *interesting* stuff: the clinical applications, the pathology, jokes and anecdotes, questions and answers. You'll notice on the Syllabus that these quizzes can be one of two formats:

> ***A Netter Quiz,*** that requires them to label ten structures on one or more Netter's plates that have been assigned in advance. Students prepare for this quiz by covering the labels on the assigned plates, and learning them.
>
> or:
>
> ***A Lecture Quiz,*** that takes questions from the Notespak for the upcoming lecture. These are short-answer fill-in-the-blank questions, true or false questions, and multiple choice questions.

Because the quizzes count for points toward their grades, the students study for them, and thereby come to class with some basic knowledge of the subject to be discussed. This allows us to be unhurried in the more interesting aspects of the day's Topic of Discussion, like pathology, anomalies, and the like.

For each topic, the lecture predates the laboratory by a week. That gives the students time to digest and further study the lecture material before going to the donor body. Nobody should dissect any tissue until they have a thorough understanding of the structures at hand.

A student may take the quiz if he or she arrives at class before the quiz is over. Each student can drop one quiz in each category, because everyone occasionally has a bad day. It's an unusual student who can get through the whole semester without stepping on a rake.

Course Information

Web Address:

Course Director: Dr. William C. Forbes

Office:

Office Hours:

Phone:

Email:

Winter Term: Monday and Wednesday afternoons

DS1 Students

Room

Lecture 1:00 – 2:30 P.M.

Anatomy Laboratory, 2:40 - 5:00 P.M.

Credit Hours: 4

Prerequisites: DBS 811 Gross Anatomy

Lecturers/Support Faculty

Lecturers:

Laboratory Instructors:

Academic Policies:

All policies in the School of Dentistry Academic Policies Handbook apply to this course, including but not limited to academic integrity, mandatory attendance, professional decorum & dress code, identification (ID) badges, preclinical and classroom decorum, use of cell phone and electronic devices, examination policies and exam/quiz absences.

Disability Support Services and Accommodations

If you need course accommodations because of a disability, if you have emergency medical information to share, or if you need special arrangements in case the building must be evacuated, please contact the Director of University Academic Services/Disability Support Services. Students with special needs are urged to identify themselves to the faculty to discuss their concerns. However, faculty are unable to provide disability accommodations without official notification from the Disability Support Services office.

Student Evaluation of Instruction

Student feedback is valued by the faculty and the administration. All students are required to complete the School of Dentistry's on-line course evaluation by a specified date. Failure to comply by posted deadline dates will result in the receipt of an F (Failing) grade of record for the Evaluation Responsibility Course. Only constructive, professional recommendations will be reported and considered.

Course Description

Purpose of the course:

The purpose of the course is to provide the foundation knowledge in the anatomy of the head and neck that will be vital to the dental curriculum, and to the practice of dentistry.

Course Goals:

This course will introduce the student to anatomical structures of the head & neck, and will provide hands-on experience with manipulation of human tissue. The course will introduce and study:

- ☐ Anatomical nomenclature

- ☐ Essentials of human anatomy of the head and neck
- ☐ Anatomical variation and how it applies to structures of the head and neck.
- ☐ Clinical applications of the study of head and neck anatomy by dental professionals
- ☐ Anatomical considerations for patient examination and dental radiology of the head and neck region.
- ☐ Locating and identifying the regions and associated surface landmarks of the head and neck on a diagram and on a patient, and integrating the knowledge of those landmarks into the physical examination and clinical practice.

Course Objectives:

At the conclusion of the course, the dental student will be able to:

- ☐ Locate and identify the bones of the head and neck and their landmarks on a diagram, a skull and a patient. Discuss certain abnormalities of bone.
- ☐ Locate and identify position of the muscles of the head and neck region on a diagram, a skull, and a patient. State the innervation, and in some cases the blood supply, origin and insertion, of those muscles.
- ☐ Discuss the processes of mastication, speech and swallowing in regard to anatomical considerations.
- ☐ Locate and identify the specific anatomical landmarks of the temporomandibular joint on a diagram, a a skull and a patient. Describe the movements and disorders of this joint.
- ☐ Trace the routes of the blood vessels of the head and neck on a diagram, a skull and a patient. Integrate this knowledge into clinical practice, especially involving problems of local anesthesia injections, the spread of dental infections or path of malignant cells by the blood.
- ☐ Identify all glandular tissue and associated structures of head and neck region.
- ☐ Identify and trace the route of the cranial and spinal nerves on a diagram, a skull or a patient. Discuss the function of each cranial nerve. Describe in detail the

autonomic nervous systems of the head and neck. Integrate this knowledge into clinical situations: the functioning of the region's muscles, joints and glands; the concepts of pain management with local anesthesia, and related disorders.

☐ Locate and identify the groups of lymph nodes of the head and neck on a diagram and on a patient. Identify patterns of lymph drainage. Integrate this knowledge into clinical practice, especially involving physical examination of a patient, and the spread of dental infections or path of malignant cells by lymphatic channels.

☐ Locate and identify the fascia and major fascial spaces of the head and neck, and communication between the various spaces. Integrate that knowledge into applications in clinical practice.

☐ Integrate the knowledge of the structures of the head and neck into the clinical practice of patient examination and treatment.

Instructional Methods:

Instructional methods utilized in the course include lecture format with PowerPoint presentations, videotapes, and plenty of give and take with students. Students are strongly encouraged to bring questions and their own previous experiences to bear on these discussions. The laboratory component will include human cadaver dissections, along with the use of prosections, plastic models and charts, plastinated specimens and skeletons. Videotapes of required dissections, and laboratory written instructions will be available on Blackboard each week before the dissections are to be performed. Class discussions and examination questions will have significant case-based components.

Course Policies

Attendance is required for all lectures and laboratory sessions. If a student misses a class for any reason, this course will follow the format described in the Academic Policies Handbook, pages 40 and 75. Every lecture session will begin with a quiz. A student may take the quiz if he or she arrives at class before the quiz is over. Quizzes will be two types: Lecture Notespaks, or Netter's Atlas. At the end of the course, each student's lowest graded quiz in each category will be dropped.

Examinations will be given promptly at the time announced, and will close at the time announced, with extra time not allowed, even if the student arrives late to take the examination.

The course Blackboard site will contain announcements for the course, as well as videotapes of the dissections, written instructions and Structures Lists posted one week in advance of the dissection. Specific policies regarding conduct in the Anatomy Laboratory are listed below:

Conduct in Gross Anatomy Laboratory

1. Dissection of the human body shall be performed with dignity and respect.
2. Students assigned to a cadaver will be responsible for the proper conduct of the dissection, according to the dissection guide and faculty instruction.
3. All identifying information about the cadaver is confidential.
4. Students may observe, but may not dissect or alter those cadavers assigned to other students.
5. Human tissues, including prosections, bones and skeletons, may not be removed from the Laboratory area. Bones, skulls and plastic models are available in the library.
6. Neither the cadaver not any part thereof may be photographed or videotaped without permission.
7. Neither the cadaver not any parts thereof are to be positioned or displayed in an inappropriate , comical or obscene manner.
8. Visitors are allowed, but **only with the *prior* specific permission of the Course Director.**
9. No eating, drinking or smoking in the laboratory. No music or other such programming is allowed in the laboratory.

Dissection is Mandatory

Dissection is a privilege and is mandatory. Every student is required to dissect. Attendance may be taken. Anyone not attending laboratory and dissecting will be called to account and required to explain how they think not dissecting advances their professional education and prepares them to take responsibility for the health and lives of their future patients.

Textbook and Resource Materials

REQUIRED TEXTBOOKS:

These required textbooks will be used extensively in the course. A student who does not have access to these books will place him/herself at a substantial disadvantage.

Atlas of Human Anatomy, Netter, Frank, CIBA-GEIGY Corp.
Gray's Anatomy for Students, Drake, Richard L.; Vogl, Wayne; and Mitchell, Adam W.M. Elsevier Publishing. 2nd Ed.

RECOMMENDED TEXTBOOKS:

Color Atlas of Anatomy, 5th Ed. Rohen, Yokochi, and Lutjen-Drecoll, Lippincott, Williams & Wilkins
Stedman's Medical Dictionary, 27th ed. Philadelphia PA: Lippincott Williams & Wilkins, 2000.

USEFUL DVDs

Anatomy DVD: *Acland's DVD Atlas of Human Anatomy, DVD 4, and DVD 5*, (which have to do with the head & neck) only. Published by Lippincott, Williams and Wilkins.

USEFUL WEBSITES:

Students should consult the Blackboard website each week, for videotapes of the upcoming dissections, announcements and written directions for the lab.

OTHER RESOURCES:

Students should be aware that models of anatomical structures, skulls, disarticulated bones, and atlases are on reserve hours.

Evaluation and Grading / Grading Scale

There are 1400 possible points in this course.

If the student receives:	*the grade will be:*
1302 or more points:	A
1260-1301.9 points	A-
1218-1259.9 points	B+
1162-1217.9 points	B
1120-1161.9 points	B-
1078-1119.9 points	C+
1022-1077.9 points	C
980-1021.9 points	C-
840-979.9 points	D
less than 840 points	E

Students will be evaluated on the basis of the following:

Two midterm examinations worth 200 points each:	400
One final examination worth 200 points:	200
Two midterm laboratory exams worth 150 points each	300
One third unit lab exam worth 150 points	150
Netter's quiz average	100
Lecture quiz average	200
Participation points (includes undetermined number of lab quizzes at 10 points each)	50

The student's lowest grade in each quiz category (except lab quizzes) will be dropped. If a quiz is missed, that will be the lowest quiz grade. Subsequent missed quizzes will be scored as a zero. Lab participation points may include grades on unannounced quizzes (10 points each; no dropped quiz in this category).

Course Evaluation Methods

Quizzes

A quiz will be given at the start of every lecture session. A student may take the quiz if he or she arrives before the quiz is over.

Midterm Examination

The midterm examinations will be in multiple-choice format. The first midterm examination will cover lecture material from Topics 1-7. The second midterm examination will cover lecture material from Topics 8-16. The final examination will be cumulative.

Practical Examination

Practical examinations will consist of 33 stations, with 80 seconds at each station. Practical Examinations will not be cumulative. Practical Examination I will cover material from the beginning of the course; Practical Examination II will cover material experienced after Practical Examination I, and Practical Examination III will cover material experienced after Practical Examination II.

Final Written Examination

Final written examination will be cumulative over the entire semester, in multiple-choice format.

The Cadavers

The cause of death and age at the time of death of your donor is posted in the laboratory. Always be aware of the fact that your cadaver is the body of someone's loved one; treat it as you would wish that of a family member, to be treated in this situation. Virtually any procedure is appropriate if your motivation is to respectfully learn in order to be able to help your future patients.

The privilege of dissection is made possible by the generosity of donors and their families. It is their intent that you make full use of the opportunity to learn from their gift. Fulfill their expectations and honor their wishes by dissecting with purpose and respect.

Please Read Carefully:
The Michigan Penal Code
Act 328 (Part)

AN ACT to revise, consolidate, codify and add to the statutes relating to crimes; to define crimes and prescribe the penalties therefore; to provide for the competency of evidence at the trial of persons accused of crime; to provide immunity from prosecution for certain witnesses appearing at such trials; and to repeal certain acts and parts of acts inconsistent with or contravening any of the provisions of this act.

History: 1931, Act 328, Eff. Sept 18, 1931

The People of the State of Michigan enact:
Chapter XXVI
Dead Human Bodies

750.160 Disinterment, mutilation, defacement, or carrying away of human body.

Sec. 160. A person, not being lawfully authorized so to do, who shall willfully dig, disinter, remove, or convey away a human body, or the remains thereof, from the place where the body may be interred or deposited, or who shall knowingly aid in such disinterment, removal, or conveying away, or who shall mutilate, deface, remove, or ***carry away a portion of the dead body of a person***, whether in his charge for burial or otherwise, whenever the mutilation, defacement, removal or carrying away is not necessary in any proper operation in embalming the body or for purposes of postmortem examination, and every person accessory thereto, either before or after the fact, shall be ***guilty of a felony*** punishable by imprisonment for not more than 10 years, or by fine of not more than 5,000.00. This section shall not be construed to prohibit the digging up, disinterment, removal or carrying away for scientific purposes of the remains of prehistoric persons by representatives of established scientific institutions or societies, having the consent in writing of the owner of the land from which the remains may be disinterred, removed or carried away.

Underscoring added.

COURSE SCHEDULE

DBS 814: Gross Anatomy II (Anatomy of the Head & Neck)
LECTURES, LABORATORIES, QUIZZES AND EXAMINATIONS

[Table 1 continues on the following pages]

Please do not ask the professor to change the dates and times of examinations, because laboratory instructors have private practices, and need to be able to plan ahead.

Subject to change:

12 Netter quizzes, drop one

Winter Semester

DATE	**TOPIC OF DISCUSSION**	**LABORATORY EXERCISE** ***Grant's Dissector pages in Italics***	**NETTER QUIZ: PLATES (5th Edition)**	**LECTURE QUIZ: TOPICS**
Session 1. Wednesday January 5	**Topic # 1:** The Skull	Distribution of skulls		
Session 2. Monday January 10	**Topic # 1 (cont'd)** The Skull	Handout: The Skull	Netter Quiz # 1: plates 4, 6, 10	
Session 3. Wednesday January 12	**Topic # 1 (cont'd)** The Skull	NO LAB	Netter Quiz # 2: plates 9, 11, Names of openings, *only:* 13	**Lecture Quiz # 1: Topic # 1:** pages 1-21 lecture notes
Monday January 17	**MARTIN LUTHER KING JR. DAY. UNIVERSITY CLOSED**			
Session 4 Wednesday January 19	**Topic # 2:** Muscles Facial Expression The Face Twelve cranial nerves	Skull. This lab will finish with a Lab Quiz on the skull.	Netter Quiz # 3: plates 15, 17	
Session 5. Monday January 24	**Topic # 3:** Temporomandibular joint The cerebrospinal fluid The Meninges	surface anatomy, take skin from face and cranium *221.1-232.2 (skip 226.1-229.5)*		**Lecture # 2: Topic # 2**
Session 6. Wednesday January 26	**Topic # 3 (continued)** Temporomandibular joint The cerebrospinal fluid The Meninges	refine dissection of face *221.1-232.2* Tape on skulls exercise: Muscles of Mastication	Netter Quiz # 4: plates 101 (top only), 102, 103, 108	
Session 7. Monday, January 31	**Topic # 4:** Cavernous Sinus, Dural sinuses Facial nerve	remove calvaria and brain *232.2-238.6 (skip brainstem and 4th ventricle p 236)*	Netter Quiz # 5: 85, 86	
Session 8. Wednesday, February 2	**Topic #5:** The Eye	cranial fossa *240.5-244.2*		**Lecture Quiz # 3: Topics 3, 4**
Session 9. Monday, February 7	**Topic # 6:** Blood Vessels of the Head & Neck	review	Netter Quiz # 6: 39, 69	

DATE	TOPIC OF DISCUSSION	LABORATORY EXERCISE *Grant's Dissector pages in Italics*	NETTER QUIZ: PLATES (5th Edition)	LECTURE QUIZ: TOPICS
Session 10. Wednesday, February 9	**Topic # 7:** Posterior triangle Cranial nerve IX Cervical plexus	review		**Lecture Quiz # 4: Topics 5, 6**
Session 11. Monday, February 14	**MIDTERM EXAMINATION I**	**PRACTICAL EXAMINATION UNIT I**		
Session 12 Wednesday, February 16	**Topic # 8:** Anterior triangle	NO LAB		
Session 13 Monday, February 21	**Topic # 9:** Root of the neck Prevertebral region	posterior triangle *251.2-254.4* and anterior triangle *254.5-259.8*	Netter Quiz # 7: 2, 24, 31 (bottom only), 32 (bottom only)	
Wednesday February 23	**Department Curriculum Review Meetings NO CLASS**			
Session 14 Monday , February 28	**Topic # 10:** Cranial nerve V	anterior triangle (cont'd), root of the neck *260.1-263.6*		**Lecture Quiz # 5: Topics 8,9**
Session 15 Wednesday March 2	**Topic # 11:** Parotid gland region Anatomy of Local Anesthesia: Maxillary	parotid region *263.6-265.7*	Netter Quiz # 8: 42, 44, 45	
Session 16 Monday March 7	**Topic # 12:** Infratemporal fossa Anatomy of Local Anesthesia: Mandibular	local anesthesia lab Local Anesthesia Lab Exercise with skull and written exercise		**Lecture Quiz # 6: Topics 10, 11**
Wednesday March 9	**NO CLASS DEPARTMENT CURRICULUM REVIEW**			
March 14-18	**SPRING BREAK**			
Session 17 Monday March 21	**Topic 13:** Autonomics, Lymphatics	temporal, infratemporal regions *265.8-270.6*	Netter Quiz # 9: 104 (top only) 63	
Session 18 Wednesday March 23	**Topic 14:** The brain	infratemporal fossa (continued) *265.8-270.6*		**Lecture Quiz # 7: Topics 12, 13**
Session 19 Monday March 28	**Topic 15:** The pharynx	brain *238.6-240.4*	Netter Quiz # 10: 64 (top only), 65, 66	

10 Lecture quizzes, drop one

Undetermined number of lab quizzes drop NONE

DATE	TOPIC OF DISCUSSION	LABORATORY EXERCISE *Grant's Dissector pages in Italics*	NETTER QUIZ: PLATES (5th Edition)	LECTURE QUIZ: TOPICS
Session 22 Monday April 4	**MIDTERM EXAMINATION II**	**PRACTICAL EXAMINATION UNIT II**		
Session 23 Wednesday April 6	**Topics # 17, 18:** Pterygopalatine fossa, Oral cavity	nasal cavities *279.2-283.8*		
Session 24 Monday April 11	**Topic # 19:** Floor of mouth Submandibular triangle	mouth, tongue *288.6-291.4*		**Lecture Quiz # 9: Topic 17, 18**
Session 25 Wednesday April 13	**Topic # 20:** Fascial planes, Spread of infection	Continue mouth, tongue. Clinical procedures on cadavers	Netter Quiz # 11: 58 (top), 59 (top), 60 (top), 61 (top)	
Session 26 Monday April 18	**Topic # 21:** The larynx	larynx, middle ear Use models	Netter Quiz # 12: 77, 78, 80 (top only), 52, 53	
Session 27 Wednesday April 20	**Topic # 22:** Review	review		**Lecture Quiz # 10: Topics 19, 20, 21**
Session 28 Date & time To Be Announced	**FINAL EXAMINATION**	**PRACTICAL EXAMINATION III**		

APPENDIX II: Lecture Notes for "Introduction to Gross Anatomy" Lecture

These are my notes for the introductory lecture in my Anatomy course.

Greetings:

1. Course name and number, my name, location of my study, phone, email address on board. The door to my study is always open to students.

2. Go over the Syllabus

3. The general outline of your learning process:

a. Every session of lecture will start with a 5-minute quiz. You will be allowed to take the quiz if you arrive before it is finished. Those quizzes are listed in the Syllabus, and will take alternating forms:

1) You will be asked to label ten structures on a particular *Netter* plate that has been assigned in the Syllabus,

or:

2) you will fill in the blanks on ten questions pertaining to the Notespak for the upcoming lecture.

b. After the 5-minute quiz, a lecture will be presented on the topic listed on the Syllabus for that date.

c. The topic of that lecture material will be dissected in the Anatomy Lab *one week later*, after you've had time to digest the information.

LAB PORTION OF THE COURSE

1. These bodies have been donated to science, either by the person himself, or by their families. We refer to them as "cadavers," "donors," or "bodies."

2. Review of cadaver procurement. Cadaver protocols:

    a. Demonstrate complete respect: no disparaging remarks, no practical jokes, nothing that's unnecessary. This donor body belonged to somebody's mother, or father.

    b. Cadavers must be kept moist. Drying out is not acceptable.

    c. Do not discard human tissue in wastebaskets. Rather, put it into the red bucket. This is important: only human tissue goes in the red buckets.

    d. Leave the plastic covering on the head until the class is instructed to unwrap it.

    e. Turn the cadaver into the *supine* position: *ventral* side up, *dorsal* side down. Get plenty of help, including at least one instructor the first time.

    f. Make a toe tag, and another ear tag, that matches the ear tag that was attached to the ear of your cadaver when it came to the lab.

3. Do not eat not drink in lab.

4. No photography is allowed in the Laboratory, except that designated by an instructor.

5. No visitors are allowed in the Lab without permission.

6. Do not remove anything from the Lab, including human tissue.

7. You MUST dispose of sharps *only* in sharps container.

8. Identification tags must be left on. *Never* remove identification tags from cadavers.

9. Select a locker for your stuff.

10. Take a new smock and write your name over your left second intercostal space. If you don't know where that is, look it up. Don't write on your smock while wearing it; the ink can carry through onto your scrubs underneath.

11. Your cadaver station will be assigned to you. The cadaver table groups are assigned, but they are assigned randomly. If one of your tablemates turns out to be your ex-fiancée, we can make a change next semester.

12. You may think your grades are not important, as long as you pass. If you think that, you are *so wrong*.

13. We take ten points off for complaining about the smell in the lab.

14. Cheating: I will see that you're kicked out of school if you cheat. I will tell the dean the following: it's the student, or me. One of us is leaving.

15. You should become familiar with the use of mnemonics. If your parents are practicing a health profession, say to them: "On old Olympus towering tops…" See how they respond*.

16. Attendance in Anatomy Lab is *mandatory*. Please do not force us to adopt Mickey Mouse procedures to enforce this. You cannot learn adequately the laboratory material from an Atlas, not from the internet.

17. Lab exams consist of 50 stations. You will have specific information beforehand about what you need to learn for the practical examination.

18. Learn the material *before* you come to Lab. This is a self-directed course. We don't have enough lab instructors to teach you everything. Come to Lab prepared, and you will love the anatomy lab.Goof off, and you'll find it entirely too challenging.

19. At the end of this course, we will have a Service of Appreciation, expressing our gratitude for the donor's generosity in furthering our education. Details will follow.

20. Information sheets about the individual cadavers are posted in the Lab. They will tell you the person's age at the time of death, and cause of death.

21. Dress properly in lab: protective eyewear, gloves, gowns are provided for you. Some courses require medical masks. When you move from one cadaver to another, change your gloves.

22. Get an instructor to demonstrate how to place, and replace, the scalpel blade. If you don't know what you're doing, bad things can happen. Ask your professor, privately, about his own maladroit experience in this regard. You can prompt him by simply saying "nine stitches in your index finger."

23. Ventilation in lab turns the over every few minutes. There are strict guidelines about ventilation in this Lab, and they are followed to the letter.

24. Cleanup policy: tidiness is expected and appreciated. You're not required to wash your dissection instruments between lab sessions, but you may, if you wish.

25. Each dissection table is equipped with a "dirty" Lab Manual. Lab Handouts, which contain additional directions more specific to this course, are distributed in advance either on the computer monitor or hard copy.

26. Make use of the bones, skeletons and plastic models that are provided. This is more 3-dimensional learning, which is the best kind.

27. We think the information presented in this course *is vital* to the practice of your health profession. We will expect you to master it before we allow you to move on.

28. Have fun in Anatomy Lab. Learn as much as you can, in this once-in-a-lifetime experience.

LECTURE PORTION OF THE COURSE

1. The study of Anatomy is primarily a study of three things: *vocabulary*, *nerves* and *relationships between structures.*
2. Cheating policy: zero tolerance. Just like in the Lab.
3. Learn well the cross section of spinal cord: spinal roots, nerves, rami
   Dorsal rami do not do much: sensory to back and motor to deep muscles of the back. Ventral rami form plexuses and intercostal nerves.
4. Blood vessels: learn the primary pattern of the arteries: ascending aorta, common carotids, subclavian arteries. Etymology will help you: break the words down into their component parts, for examples:

   carotid: deep sleep

   brachial: arm

   cephalic: head

The veins will not be stressed in this course, but you must learn the basic pattern: the venae cavae and so forth.

5. Vertebrae: 7/12/5. What time do you eat breakfast? Lunch? Dinner?
6. Appendicular vs. axial skeleton
7. Anatomical position
8. Origin, insertion, action and innervation of muscles
9. The skull: its bumps and foramina
10. There are five sources for questions on written examinations:

    a. handouts distributed in class

    b. anything written on the whiteboard during lecture

    c. particular, assigned in advance, atlas plates

    d. any information on lab handouts

www.ingramcontent.com/pod-product-compliance
Lightning Source LLC
LaVergne TN
LVHW071243210926
848486LV00007B/332
* 9 7 9 8 8 8 9 2 5 3 9 3 8 *